GRAVITY

How Gravity is Created

**A THEORY FOR THE NATURAL CREATION
OF GRAVITATIONAL FORCE (GRAVITY)
CAUSED BY THE PARTIAL AND DIFFERENTIAL
ABSORPTION OF OPPOSING GRAVITATIONAL
WAVES AS THEY PASS THROUGH MATTER.**

**ALSO PROVIDING AN EXPLANATION FOR
THE PHENOMENA OF INERTIA AND MOMENTUM
AND DISCUSSING AN ALTERNATIVE EQUATION FOR
"NEWTON'S UNIVERSAL LAW OF GRAVITATION".**

Dr Peter Roberts BSc MSc PhD

*"Gravity is the acceleration response created in matter
when gravitational waves pass through it.
Gravity does not exist outside of matter; it is not a force linking objects."*

*"EMF waves are the waste by-product emitted by the stars. They have no purpose.
Gravitational waves, on the contrary, weave the essential fabric that holds our universe
together and keeps it working."*

Peter Roberts 2018

SOME UNSOLICITED TESTIMONIALS

"There is probably no natural phenomenon more fascinating and directly relevant to humanity than gravity. Attempts to explain gravity predate the important contributions made by Galileo Galilei, Robert Hooke and mathematical physicist Isaac Newton. However, none of those nor earlier investigators were able to shed light on the underlying cause. Many, certainly not all, astrophysicists now subscribe to Albert Einstein's general relativity theory, but beyond that theory's complex 4-D space-time mathematics there remains wanting the scientific basis for gravitation, an explanation for what exactly draws an apple from the tree toward the center of earth. When the basis for gravitational phenomena is finally known, it will undoubtedly serve scientific and technological advances as no other knowledge advance has.
"GRAVITY - How Gravity is Created" by Dr. Peter Roberts presents a refreshingly new look at the problem, supported by novel and rational explanations ranging from earth to cosmic science.
"This book is well written, amply illustrated and well worth reading - Rodney Savidge, PhD, Principal Investigator of the 2009 APEX-Cambium experiment on the International Space Station (ISS)."

"I do think you have in this book a very excitingly new perspective that, of course, is likely to be firstly ignored, then soundly criticized and challenged, but I think it may just be able to withstand those tests. [R.A.S.]

I have reason to believe your theory is credible as well as provable and should be found in the library of every research facility and university where scientists and students are most prone to reading without interruption. (D.W.]

Thank you for the opportunity to read this remarkable work. [H.G.K.]

Newton was reported to have stated that his work was relevant only because he could stand on the shoulders of past giants. Your work is, of course, a step beyond. [H.G.K.]

Congratulations on your publication. We know it was hard work. We are certain that it will contribute to the advancement of the understanding of this great phenomenon of nature and to the development of new technologies. [D.L.M.] Member of The Gravity Club.

I know that your book will be of interest to the general public [G.D.]

I think it's perfect! [P.B.]

I believe you have an effective presentation suitable for any college level student to understand. [D.W.]

This book is really well explained with fine pictures and easy-to-understand vocabulary. I couldn't see any possible improvement. Nice Work. [S.P.]

I like the idea of your proposal as it makes more sense... [W.N.]

What I like the most about your book is your strict logical reasoning. The mechanism of gravitational waves is nicely explained and it fits the observations. It is a good read! [P.T.]

It would be a great textbook for your class. [C.W.]

There is no doubt that because it is so well written and full of interest, it should be well received by the general public. [G.N.]

I hope that your thinking gathers traction, acting as a catalyst in bettering understanding for mankind. [M.D.]

I am half way through your theory and I am enjoying it very much. Finally we may just have a breakthrough that is so much needed. You have dispelled spacetime which is monumental in my view, but think about it—do they really deserve you yet ? No! I would just like to say I magnificently admire your valiant approach and will wholeheartedly follow you whatever you need. You are a hero in my eyes. [A.L.]

Thank you for the book, it really is the new angle of thinking. I think most importantly it should be available to all the readers, especially to students who would think and develop on it more. [N.S.]

I do accept the tenet that the gravitational field creates a drag effect. Not to question whether the effect is real or not, I accept it as a pillar of your work. Your theory is very fertile ground for expanding backwards into field theory. Your principle of drag effect may point to a way of reconciling the trinity of electric, magnetic and gravitational theory. I agree 100% with your observation about Einstein, who understood the limitations of his theories and knew he was plugging in fudge factors to make it work, but then that all got forgotten as the rest became "settled science". [M.P.S.]

Such a great theory. [M.P.S.]

I have no problem with your theory of gravity - it's an interesting hypothesis, which could well be true. [P.S.]

It's clear that you are entirely persuaded about the correctness of your ideas, and frankly I think they're brilliant and deserve very serious consideration. [R.A.S.]

I am very interested in how you, a civil engineer with an apparent focus on geotechnology, as well as a side interest in languages, came to create a theory that to me, a lowly engineering student, seems to be the Ockham's Razor for modern physics? [J.C.B.]

It is well written and easy to follow, proposing a blindingly obvious and satisfyingly unifying theory, which explains a lot more than past theories of gravitation. (Bear in mind that my physics stops at first year BSc). I really like the way that you can so easily explain the basic principles so early in the book. Including the concept of why there is so much omnidirectional background gravitational wave flux. [N.C.]

Miscellaneous notes from the author:

Please note that the author's theory is primarily to explain how gravitational force is created as a challenge to the concept of spacetime. He does NOT make any attack on time dilatancy, bending of light, special relativity, or general relativity, all of which he looks upon as part of the overall mathematical explanation for how matter *behaves* in response to gravity, but not how gravity is *created*.

The author proposes how gravitational force is created. (What nature does with gravitational force, he leaves up to Einstein and nature.)

A note for students: This book may be used for exam purposes as an example of a new theory which proposes that the creation of gravitational force may take place in three-dimensional space but with effects that may still be described in four-dimensional relativity theory. The author's theory does not purport to replace the Einstein concept of relativity but adds a new gravitational foundation to it and, in the process, provides a new three-dimensional explanation for inertia and momentum.

A note for teachers and professors: It is good for high-achieving pupils and students to be able to present alternative 'solutions' in exams for purposes of comparing and contrasting. This practice demonstrates extensive reading and shows an understanding of the subject in depth. It is good academic practice for students to learn open-mindedness in their study of science in order that they may benefit from the varied perspectives offered by alternative systems and theories.

A note for professional scientists and persons with advanced knowledge: This book uses the word 'atom' in order to simplify the sense of the theory and to not become involved in an irrelevant discussion on the general subject of quantum gravity and field theory. The author is of the opinion that Quantum Field Theory is likely to be the one that moves forward most successfully into the future. However, it is immaterial whether the author's theory considers that gravitational acceleration is imposed on a constituent atom or on a constituent quantum field. The author has read Einstein's own comments on the gravitational field and field theory and finds them capable of being criticised in places. This book is not the place to discuss those aspects because Einstein was unable to explain much of what he wrote and upon which he based his theories of relativity; therefore, direct comparisons cannot be made.

Definitions: In this book, the terms "gravitational force" and "gravity" are synonymous.

Peter Roberts

The author holds three British university degrees: BSc (Geology), MSc (Mining & Tunnelling), and PhD (Civil Engineering). He is now retired but had achieved, before his retirement, 25 years of Fellowship of the UK Institution of Civil Engineers, and over 35 years as a Chartered Engineer. His qualifications and former affiliation titles include:
Professor BSc MSc PhD CEng CGeol CText FICE FIMMM FIGeol FTI FGS.

Rp

Russet Publishing
russetpublishing.com

First Printing: 2018
v1.1

ISBN   978-1-910537-34-3

Russet Publishing
United Kingdom
www.russetpublishing.com

Peter Roberts would welcome personal communications to
peter.roberts970@gmail.com

To order this book from Lulu and to see other works by Peter Roberts go to
lulu.com/spotlight/PRR

His books can also be purchased from amazon.com, amazon.co.uk, Book Depository, Barnes & Noble, Waterstones, and many other retail distributors worldwide.

Special discounts are available on quantity purchases at Lulu from 15 and upwards.

For other enquiries, please contact Russet Publishing:
editor@russetpublishing.com

This printed book has also been produced in PDF electronic format at A4 size with a significantly reduced price. Additionally, there are electronic iBook, ePub and Kindle formats. Electronic versions of textbooks sometimes have slightly less-sharp diagrams to keep their file size down, and they provide a less easy reading environment if used on small tablets or phones.

Printed and distributed internationally by Lulu Press Inc.,
Raleigh, North Carolina, USA.
lulu.com

Dedication

I dedicate this book to all the young scientists who have been brought up and trained to believe in Einstein's spacetime but who are bright enough to have questions about the concept.

To those who do not have closed minds, and who recognise that we have experienced three hundred years of amazing progress in every field of technology except that of creating, controlling, and harnessing gravity for the benefit of humanity.

To those engineers, scientists, and researchers who wonder why that might be.

To those people who do not think that the subject of gravitation can only be considered by advanced mathematicians and astronomical physicists.

Also, to those who simply have a great interest in the most important challenge facing humanity at the present time—gravity.

To you, I dedicate this book containing my new theory of how the force of gravity is created.

I wish you all well and hope that my book will help you to realise that there are other ways of looking at the subject of gravity which may make the breakthrough that we so urgently need.

Peter Roberts.

Acknowledgments

I thank my friends at The Gravity Club on Facebook and all the members of Quora who have helped me by contributing their useful and much-appreciated comments.

In particular, I thank the friend who did the final proof-read of the full book—a great deal of work which, undoubtedly, benefitted the book and for which I am most grateful. Thank you.

Finally, of course, I thank my wife for her support and interest while I wrote this book.

CONTENTS LISTING

Preamble

This book may, perhaps, allow scientists to alter their views of how the three-dimensional universe is constructed and functions. The author hopes that it may form the basis for reconsideration of those subjects as, herein, he lays out a comprehensive concept (for the first time) of a conventional physical mechanism that creates gravitational force.

To the author's best knowledge, the theory of differential opposing partial absorption (DOPA) of gravitational waves by matter has never been proposed before. He puts it forward for consideration as a new concept for science.

Roberts' theory of 'Differential Opposing Partial Absorption' of gravitational waves is referred to hereafter as either 'DOPA' or 'absorption' theory.

This book provides a comprehensive review of how gravitational force is created and how potential gravity zones are formed around cosmic bodies—a process presently attributed to the existence of four-dimensional spacetime.

The greatest scientific challenge existing today and the most important scientific challenge for humankind is not the esoteric study of far distant astronomical bodies, nor the microscopic analysis of the constituent particles of matter at CERN, but the study of how the force of gravity is created, and how that force interacts with and applies itself to matter.

Because no one has, so far, been able to suggest how gravity is created in nature and how it applies itself to matter, it has been impossible to research the subject. And yet, the potential outcome of such work could change civilisation more than all of the scientific advances made so far over the last ten thousand years.

The theory presented in this book does not contradict either Newton or Einstein where their theories involve the prediction of the behaviour of matter in relation to the force of gravity. It newly proposes a mechanism for the production of gravitational force that was absent in both of their works. It addresses the problem that both Newton and Einstein experienced in trying to explain how gravitational force is created and how it applies that force to matter.

Newton claimed that every particle of matter was in a permanent state of attraction to every other particle of matter in the universe simultaneously. He was unable to propose or justify how this attraction was created or implemented. He was obliged to imply that this force acted faster than the speed of light, in order to avoid time lag problems over billions of light years of distance. In doing this, he contravened what are now Einstein's presently-accepted theories of relativity. Of course, he did not know that at the time!

Einstein claimed that there was such a thing as spacetime in which time was an intrinsic fourth dimension. He claimed that the very presence of matter within four-dimensional spacetime somehow distorted or warped it. He was unable to propose why or how matter did this. Einstein further proposed that once matter had distorted spacetime, other matter in its vicinity was obliged to travel along that curved distortion, thus accounting for the apparent attraction proposed initially by Newton. Einstein, therefore, claimed that gravity was not an attraction but a response to spacetime warping. He was unable to justify why spacetime exists and, further, was unable to explain how matter was supposed to distort spacetime. The second problem with Einstein's spacetime hypothesis was the same as that experienced by Newton; an inability to account for how his single four-dimensional field transmitted information over considerable distances. This was a significant weakness in the construction of his otherwise-mathematically-sound relativity theories and stems from their being built on a false spacetime foundation. Relativity is a wonderful edifice built on shifting sand.

That did not mean that relativity does not work. It works very well, but a practical reason for the existence of potential gravity zones (previously called 'gravity wells') around objects has, heretofore, not been proposed.

The theory described in this book proposes and explains the four things that neither Newton nor Einstein could: 1) the nature of the mechanism that creates gravitational force, 2) how that force is applied within objects, 3) how that force is applied between objects and 4) why a working mechanism for gravity does not need light-speed transmission of information. (DOPA theory meets all four of these criteria.)

The author's theory not only meets, but explains all four requirements. To the best of the author's knowledge, his theory is the first to do this.

Finally, therefore, what the author is proposing is that we can now create a link between the three-dimensional world in which we live and the four-dimensional description of how matter behaves according to relativity.

Peter Roberts

November 2018

SUMMARY

This book demonstrates, for the first time, that a theory for 'how gravitational force is created in nature' can be set out in simple, non-relativistic terms, and whose principles match all the observations that can be proposed and made for its testing. At least it meets all tests so far, as described within this book. This theory is independent of any aspects of relativity, its physics being straightforward and requiring no advanced mathematics to comprehend.

DOPA theory provides the long-sought-after mechanism for the creation of the universal gravitational field. Einstein, in his theories of relativity, Chapter 9, entitled, 'The Gravitational Field', wrote, *"The Earth produces in its surroundings a gravitational field..."* However, read on as one might, one never finds any explanation for how that gravitational field is created, nor how it can exist concurrently with 'warped spacetime'.

Subsequently, in the same chapter, Einstein wrote, *"...the gravitational field exhibits a most remarkable property, which is of fundamental importance. Bodies which are moving under the sole influence of a gravitational field receive an acceleration which does not in the least depend either on the constituent material or the physical state of the body."* Again, Einstein does not provide any explanation for how this comes about, but absorption theory does. It explains precisely why a small body falls at the same accelerating rate as a denser large body of greater mass and density. No one has been able to do this up to this point. No one, prior to reading DOPA theory for the first time, is able to explain that mechanism—whether a lay-person or an astrophysicist. Having read this theory, they will be able to do so. That is unique and demonstrates the novelty and validity of Roberts' work.

Among other things, this book describes the author's prime tenet that gravity waves are ubiquitous, forming an omnidirectional flux (which is the potential gravitational field), at least in the greater, non-peripheral part of the universe. The theory discusses how the increasingly-radial mono-directionality of the wave flux towards the edges of the universe may account for the outward acceleration of matter in those regions.

It is proposed that gravitational waves pass through matter with ease, but that as they pass through, they lose some of their energy by imparting a minuscule physical force, in the direction of wave travel, on each atom encountered. The physical effect of gravitational waves on matter has been demonstrated by the 2015-2017 LIGO recording and measurement of gravitational waves for the first time. Roberts considers that the LIGO results confirm that gravitational waves are different from EMF waves. *(LIGO is an acronym for the Caltech/MIT Laser Interferometer Gravitational-wave Observatories at Livingston and Hanford, USA.) (EMF is an acronym for Electro-Magnetic Field.)*

This new theory proposes and explains how the differential partial absorption of opposing gravitational wave energy by matter produces the force of gravity that causes atoms and dust particles to coalesce and form stars and planets and which forms them into spheres and holds them together.

Absorption theory proposes that, at a planet or star's surface, the radiation outward of transiting, depleted gravitational waves (having passed through the planet's mass), will always be directly opposed by un-depleted, full-strength, incoming gravitational waves. The incoming gravitational waves, reduced by the depleted outgoing waves, produce a net inwards gravitational force that we call 'gravity'. The more massive the planet, the more depleted the outward waves are and the greater the net inward gravity. This theory, therefore, explains why surface gravity is proportional to the mass of any planet or star, its density, and inversely to its radius. Additionally, it explains that the force of gravity can only exist within matter that is being penetrated by gravitational waves.

Above and beyond any planet's surface, the outward-travelling depleted waves mix with incoming pristine (un-depleted) waves to create a 3-dimensional zone of varying *potential* gravitational force which is only activated/utilised when matter is present within it, whether that matter be a star, a planet, a moon, a comet, a dust speck, or an atom.

Furthermore, this theory uniquely describes the existence of the active gravitational force field *within* the matter where the force is generated, as a direct product of the force-creation process. The proposed mechanism for the production of gravitational force—as described in Chapter 3—explains why that force decreases to zero towards the centre of any planet or star. DOPA theory operates consistently as a working mechanism from the centre of any planet to the interplanetary space outside it and, in passing, provides a rational mechanism for shell theory. Neither Newton's attraction theory nor Einstein's spacetime theory does this.

It is the 'potential gravitation' field that creates 'gravity zones' that are only felt as such by matter itself through the proposed mechanism. This theory explains, for the first time, why matter moves towards other matter or tries to—a force which can, naturally, be resisted by orbital centrifugal action. The theory's partial absorption mechanism creates a 'shadow' zone of reduced gravitational wave amplitude between any two or more bodies, altering the local field properties, and resulting in a net acceleration towards one another. This vectored acceleration is not a connection between the bodies but is a tendency for each object to move towards the other, consequent upon the intervening reduced gravitational waves' amplitudinal strengths.

Roberts' potential gravitational field is the three-dimensional equivalent of Einstein's four-dimensional 'spacetime', but is pragmatic (not requiring the invention and acceptance of a four-dimensional version of reality) and is—obviously—more realistic, avoiding the problem of spacetime's connectivity which must somehow progress at the speed of light. In absorption theory, gravitational force is created 'on the spot' from the pre-existing gravitational wave field. The Earth is not connected to the Sun or Moon, nor to its nearest stellar neighbour; it simply behaves as 'instructed' by the pre-existing potential gravitational wave flux. There is no force between objects pulling them together. Their internally-induced and independently-induced gravitational forces automatically try to move them towards one another.

This theory also accounts for the development of a common centre of gravity between binary orbiting objects such as the Earth and Moon. It explains the fundamental mechanism for the development of high ocean tides on both sides of the planet for the first time, answering the question of why two bodies connected only by 'gravitational force' should develop a common centre of gravity. Without a realistic theory of how gravitational force is created, there can be no reliable answer to that question.

The DOPA principles, as explained in this book, conform to both 'Newton's shell theorem' and the creation of the 'Roche limit'. *(Insert those phrases into Wikipedia for explanation.)* DOPA theory conditionally emulates Newton's inverse square law for the spherical dissipation of gravitational force away from any given object. These and the other 'tests' of the validity of the theory are considered important in that none have been examined that contradict the author's differential absorption theory as propounded herein.

It is, further, proposed—as a remarkable and unexpected by-product of this work— that the partial absorption theory accounts for the creation of inertia and momentum, both of which the theory claims, and explains, are the same thing when viewed from two different reference points. It is believed that this theory is the first to explain the mechanism for the creation of inertia and momentum and their direct proportionality to mass, thus cutting across the old divide of 'gravitational mass' and 'inertial mass'. They are both shown to be the same thing.

This book illustrates the many physical circumstances that absorption theory supports. It is proposed that this theory can, and should, be accepted as a three-dimensional mechanism that can be recognised independently but whose effects can be described as part of Einstein's representation of how reality operates using the concept of relativity. This approach can open up new fields of research into what the author considers to be the most important current field of science.

The author's theory certainly withstands inspection favourably when compared with the unacceptable invention of Newton's infinite and instantaneous attraction of atoms across the universe, or when compared with Einstein's adoption of an infinite four-dimensional spacetime across the universe which is inexplicably distorted by the presence of matter.

PAGES FOR WRITING NOTES

When you are reading this book, if you spot any errors, or have any point that you
think would be useful for yourself or other future readers, please write them here, so
that a reader may benefit from them before starting to read.

_____

_____

_____

_____

_____

_____

_____

_____

_____

_____

_____

_____

_____

_____

_____

_____

_____

When you are reading this book, if you spot any errors, or have any point that you
think would be useful for yourself or other future readers, please write them here, so
that a reader may benefit from them before starting to read.

— —

— —

— —

— —

— —

— —

— —

— —

— —

— —

— —

— —

— —

— —

— —

— —

— —

CHAPTER 1 - THE PRINCIPLE AND PURPOSE OF DECLARING TENETS.

A tenet is a fundamental assumption upon which reasoning and propositions are based.

The purpose of declaring tenets in an academic proposal such as this is to lay out transparently what assumptions the author has made in order to come to his academic or theoretical conclusions. This procedure allows the reader to consider in advance whether or not he/she agrees with the basis of the proposals before entering into reading the entire work.

Such an approach enhances the reader's appreciation of the scientific proposals being made. It is possible, for example, that a reader might agree with most of the presumptions and could agree to bear with one in order to see whether the arguments presented can persuade him to change his mind. The main point is that the tenets (assumptions) provide a 'road map' of how the book will be laid out and where it will take the reader before setting out on his academic journey with the author.

Conversely, if a reader disagrees strongly with all of the tenets, then the author has saved them time and trouble, as the potential reader can reject the book at that early stage.

The declaration of tenets, as adopted herein by the author, is an honest and open approach to writing a book and to trying to convince others of a new conceptual proposal.

This theory has four presumptive tenets. Firstly, that gravitational waves exist. Secondly, that they are travelling through the universe in every direction. Thirdly, that they penetrate and pass through matter with great ease. And, fourthly, that in passing through matter they impose a tiny accelerating force on each atom in the direction of wave propagation, giving up energy in exchange.

None of these four tenets is far-fetched; on the contrary, each is factually supported by documentation, observation, and scientific research. If you can believe in each of these tenets separately, then you can believe in the book's conclusions overall when you have read it. So, when you have read each of the tenets as subsequently described, consider carefully whether or not you can accept it unconditionally, accept it provisionally subject to argument, or whether you reject it out of hand.

CHAPTER 2 - EXAMINATION OF THE FOUR TENETS.

2.1   TENET 1. That gravity waves exist.

When the author submitted his first paper on gravitational repulsion to the journal *New Scientist* in 1978, the existence of gravitational waves was speculative, but now, their presence has been confirmed in the 2016-2018 observations and recordings of gravity waves by the USA's Laser Interferometer Gravitational-wave Observatory (LIGO). So, tenet 1 is established as factual—see pages 6 and 16. (Note: In this book 'gravitational force' is synonymous with 'gravity'.) It is recognised that the LIGO work has been subject to criticism as a result of questionable processing of the incoming signals. The author believes that this can only be for the good in that it will lead to revised processing routines and improved reliability in the future.

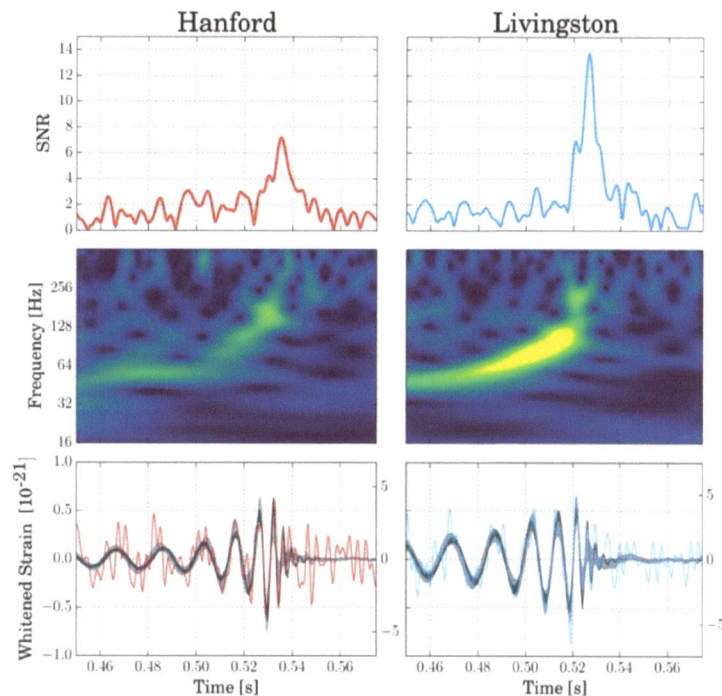

Fig 1. LIGO Gravity waves.

Figure 1 shows the gravity wave pattern received by two separate LIGO stations at Hanford and Livingstone. Note that the wave frequency is between 40 and 200 Hz with a time spread of about a tenth of a second. The gravitational wave signal arrived first at the Livingstone detector and six milliseconds later at the Hanford detector. Another six milliseconds later, it arrived at the European Gravitational Observatory's Virgo detector in Cascina, Italy.

This information was of great value in confirming the source of the waves and that the signal propagation did not exceed the speed of light.

The form of the LIGO gravitational pulse is interesting (as can be seen in the lower diagrams above) in that the signals are first received at a wavelength of about 40 Hz, increasing within a period of only six wave transits to 200 Hz. The information contained in Figure 1 needs some thought. We can calculate that the wavelength of this gravitational wave is around 7,500 km transitioning to 1,500 km or thereabouts, but, it is a part of this theory, that the wavelength of gravitational waves are proposed to be immaterial to their functionality.

These results raise three points that must be discussed in relation to absorption theory.

Firstly, gravitational waves are not EMF waves such as light or microwaves because that type of wave creates an electromagnetic response in matter through which it passes with difficulty, whereas the measured LIGO gravitational waves created a physical response to their transit through the apparatus.

Secondly, the isolated pulse nature of the waves detected might seem to challenge this theory's second primary tenet of an omnidirectional flux. Why, one might ask, if absorption theory is correct, did the LIGO apparatus not detect all of the incoming wavefronts from every direction, rather than just this one event from a particular direction?

The answer to this includes the fact that, between September 2015 and August 2017, LIGO detected seven separate, powerful, gravitational wave events, not just one. But, most importantly, the LIGO array has been in development for many years, only just reaching the stage where it can detect and record the most powerful of bursts emanating from point sources far removed in the cosmos. There can be little doubt that, in the future, as sensitivity increases, gravity wave detectors will map and plot the omnidirectional receipt of gravitational waves from every far corner of the universe.

Taking some quotations at random from Wikipedia:
"Ordinary gravitational waves' frequencies are very low and much harder to detect, while higher frequencies occur in more dramatic events and thus have become the first to be observed."
"Gravitational waves have a solid theoretical basis, founded upon the theory of relativity. They were first predicted by Einstein in 1916; although a specific consequence of general relativity, they are a common feature of all theories of gravity that obey special relativity."
https://en.wikipedia.org/wiki/Gravitational-wave_astronomy#Development
And, *"As a young area of research, gravitational-wave astronomy is still in development..."*
And *"LIGO and Virgo are currently being upgraded to their advanced configurations. Advanced LIGO began observations in 2015, detecting gravitational waves even though not having reached its design sensitivity yet; Advanced Virgo is expected to start observing in 2016."*
https://en.wikipedia.org/wiki/Gravitational-wave_astronomy#Development
And, *"...a single detector at Hanford LIGO, a gravitational-wave observatory, registered a gravitational-wave candidate occurring 2 seconds before the gamma-ray burst. This set of observations is consistent with a binary neutron star merger,[7] as evidenced by a multi-messenger transient event which was signalled by gravitational-wave, and electromagnetic (gamma-ray burst, optical, and infrared)-spectrum sightings..."* These synchronized gravity wave/EMF signals speak to and confirm that gravitational waves travel at the speed of light.
https://en.wikipedia.org/wiki/Gravitational-wave_astronomy#Development

Thirdly, whilst the reality of gravity waves has now been tentatively recorded, and whereas the author could only speculate on their existence in his 1978 monograph, it is still not certain that what has been measured by LIGO is a *direct* reading of a wave package; it might well be a modulation effect superimposed on an underlying train of much shorter-wavelength true gravitational waves. The author feels that this is not only a genuine possibility but proposes that it is a likely alternative.

Although fundamentally different from gravitational waves, it is interesting to consider that the spectrum of electromagnetic waves is incredibly broad. There are EMF gamma rays with wavelengths around one picometre—which is a millionth of a micron. A helium atom has an estimated effective diameter of 62 picometres (see "picometre" in Wikipedia). If we already know of electromagnetic waves that have a wavelength of a sixtieth of an atom's diameter, then there is nothing to suggest that shorter wavelength, higher frequency, gravitational waves do not exist that are, so far, undetected.

However, whilst speculation about that is interesting, it is irrelevant to absorption theory whether gravitational waves are as detected by LIGO or whether they are ultra-short-wave gravitational radiation because it is their properties of penetration of matter and drag creation that are cornerstones of the theory. Both of these have already been proven by the very act of generating an effective physical change of length in the LIGO apparatus reading arms. **We can conclude that gravitational waves do exist and that they exert a physical force on matter.**

Having said that, the author does not propose that these recently-measured gravitational pulse bursts are the ones that create and control the everyday gravity that we experience. That is the everyday gravity that forms and holds planets together as spheres and holds our planets in their orbits. If those things were dependent upon individual excessively-strong pulsed signals such as the LIGO results, there would be no credible way that the steady state of the universe could be maintained. What the author proposes explicitly in this book is that it is the uniform background gravitational wave flux that permits this theory to be put forward and that provides the second-by-second control of all matter within the universe. No doubt in the future and far away, as the LIGO pulses expand in their spheres, they will dissipate and become sufficiently dilute to form part of the background flux without standing out. The author only mentions and discusses the LIGO work as proof that gravity waves do now appear to exist and do interact physically with matter, upon which he could only speculate in his initial 1978 paper.

2.2   TENET 2. That gravitational waves are travelling through the universe in every direction.

An examination of the universe around us using optical, X-Ray, radio-wave, or any other telescope shows, without fail, that all types of EMF radiation reach our planet from every aspect of the universe, whether directional or temporal. But what about our newly-confirmed gravitational waves?

Wikipedia says: *"Scientists think that powerful gravitational waves are created when two extremely dense objects—like two neutron stars, two black holes, or a black hole and a neutron star—orbit* [or collide with] *one another in binary pairs."*

Fig 2. Binary neutron stars. (Artist's impression courtesy Wikipedia.)

Wikipedia says: *"There are thought to be around 100 million neutron stars in the Milky Way."*

A high proportion of these is known to comprise neutron star pairs, or neutron star/black hole pairs, or black hole pairs. There are quadrillions of galaxies in the universe that we can observe with our instruments. So, we can expect that of the order of 100 million quadrillion sources of 'event' gravity waves exist either forming the background source or in addition to the background source.

Again, the omnidirectional nature of gravitational waves might be questioned on the grounds that the LIGO waves were highly monodirectional and recorded as such. But, it is well known that the LIGO apparatus only attained the ability to record the strongest of wave pulses during the period 2016 to 2018 and is currently not capable of measuring weaker omnidirectional gravitational waves. Doubtless, further development of the sensitivity of the LIGO and other apparatuses, together with necessary modifications, will reveal gravity waves arriving from all directions equally.

The author believes that our detection of gravitational waves from binary neutron star and black hole sources and our knowledge of their number is more than reasonable evidence to support his 1978 forecast that, in our region of the universe, and throughout it, there exists an omnidirectional flux of gravitational waves that we are only just becoming able to detect and study.

To examine this in some more detail, we can start by considering how light behaves and how it is observed. The reason why light, radio waves, microwave radiation, and all other EMF radiations travel across the universe in every direction is that their sources are incredibly numerous and are located everywhere—just as for gravitational waves.

The following images reveal the scale of those quintillions of sources. (Courtesy Google.)

6

**This is a view of a galaxy such as our Milky Way.
200,000 light years across,
containing some 200,000,000,000 stars.**

Fig 3. A galaxy such as the Milky Way.

**This is a view of far distant galaxies seen in
standard optical light.
Image width 2 million light years**

Fig 4. Multiple galaxies.

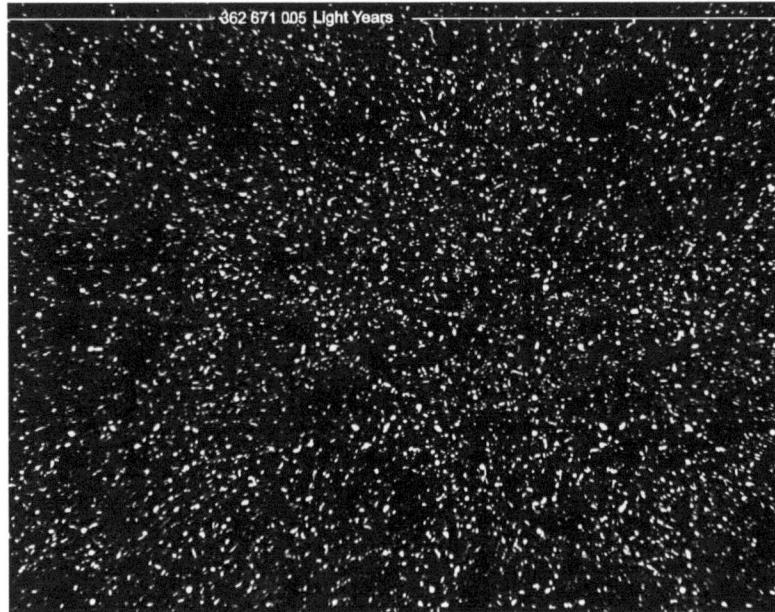

Each dot is a galaxy containing 200 billion stars, shown on a field 360 million light years across.

Fig 5. Galaxies, like stars in the sky.

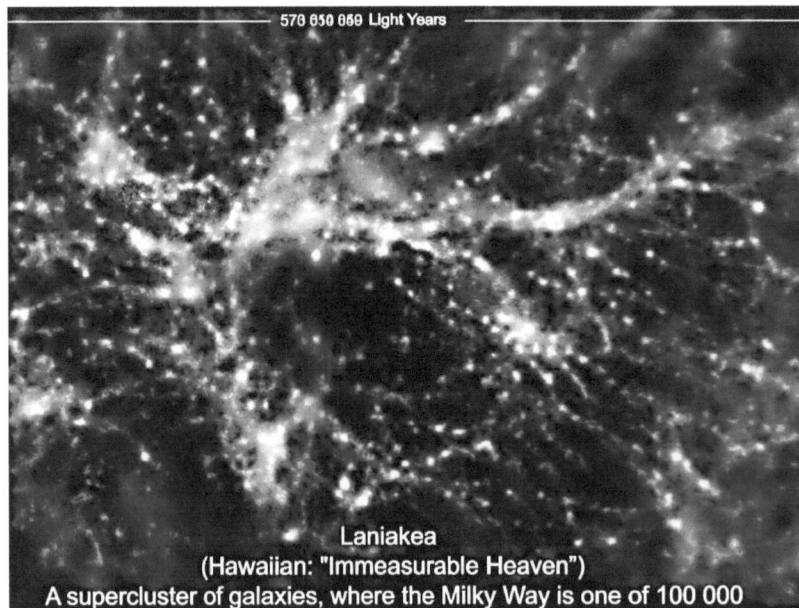

Laniakea
(Hawaiian: "Immeasurable Heaven")
A supercluster of galaxies, where the Milky Way is one of 100 000

On a field 570 million light years across, the galaxies become a cobweb of light, with each galaxy no longer distinguishable.

Fig 6. Galactic cobweb.

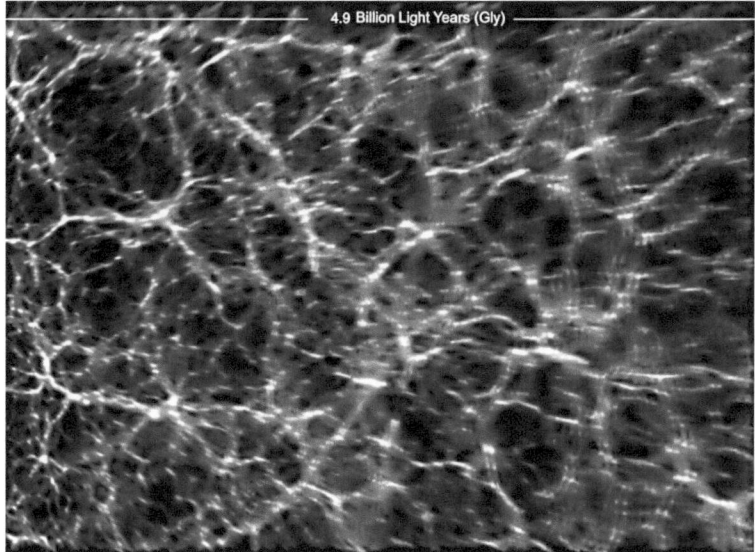

The webs of galaxies now shrink to become a soft-glowing pattern when viewed on a field some 4,900 million light years across.

Fig 7. A glowing pattern.

The cosmic web that we now see contains many billions of galaxies spread across a field 7,700 million light years across.

Fig 8. Cosmic web.

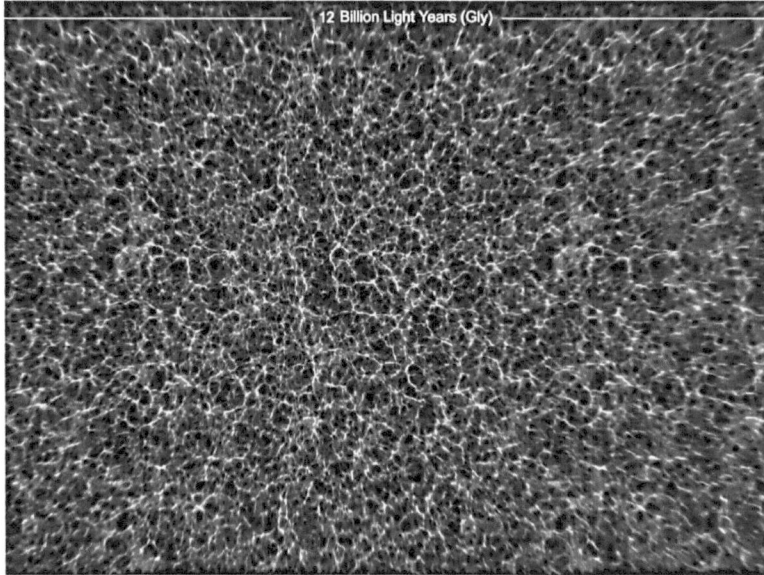

At 12,000 million light years across,
the hundreds of billions of galaxies
become a delicate lace network.

Fig 9. Lace network.

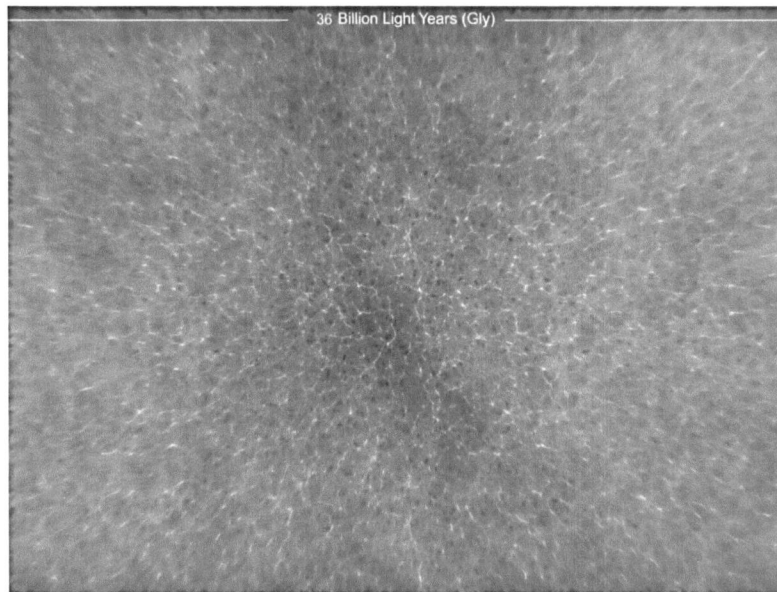

Finally, in a view spanning 36 billion light years,
our universe becomes a mist of many trillions of
galaxies and thousands of quintillions of stars.

Fig 10. Galactic mist.

It is difficult to avoid the conclusion that, in a universe that is over thirteen billion years old, and with the density of galaxies shown in the above photographs, in the same way that EMF radiation is established everywhere, gravity waves are also established throughout the universe and are travelling across it in every direction, unabated.

That confirms Roberts' theory's second tenet—gravity waves fill the universe with an omnidirectional flux.

Below, Fig. 11 is of a computer graphics video film showing the range of scale involved in our study of gravitation. It is amazing. It is impossible to demonstrate this on the printed page or even in electronic publications, but the author recommends most strongly that you go to the following URL and view this video before proceeding onwards into this theory. The object of showing it to you is to emphasise the immense nature of the universe as a source of EMF and Gravitational Waves.

https://www.youtube.com/watch?v=jfSNxVqprvM

THIS VIDEO SEQUENCE WILL CONVINCE YOU OF THE SCALE OF THE UNIVERSE, AND THE UNCOUNTABLE NUMBER OF SOURCES THAT EXIST FOR THE CREATION AND EMISSION OF EMF AND GRAVITY WAVES.

You are looking for the image of Louise, courtesy of Google.

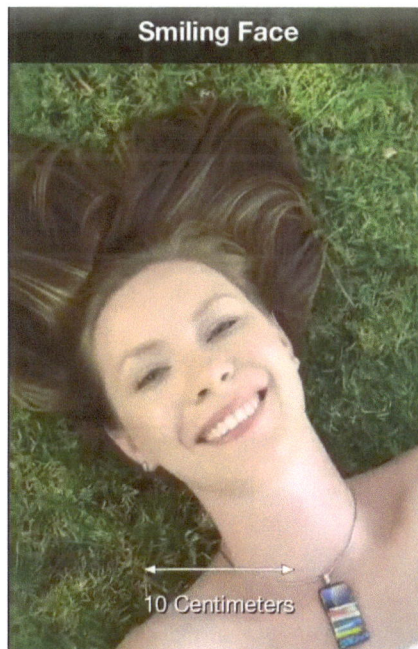

Fig 11. 'Louise'. Copyright Google and Europa Technologies

This video will take you on an incredible journey from our normal world to the far distant edges of both our huge and microscopic universes. And if you love that one, here is another URL of the very latest visualisations of how the stars are moving in our own galaxy. This video uses GAIA data only retrieved since January 2014 and synthesised much more recently. Wonderful! *(Gaia is a space observatory of the European Space Agency designed for astrometry: measuring the positions and distances of stars with unprecedented precision.)*

https://www.youtube.com/watch?v=LOJ1XmbSKhM

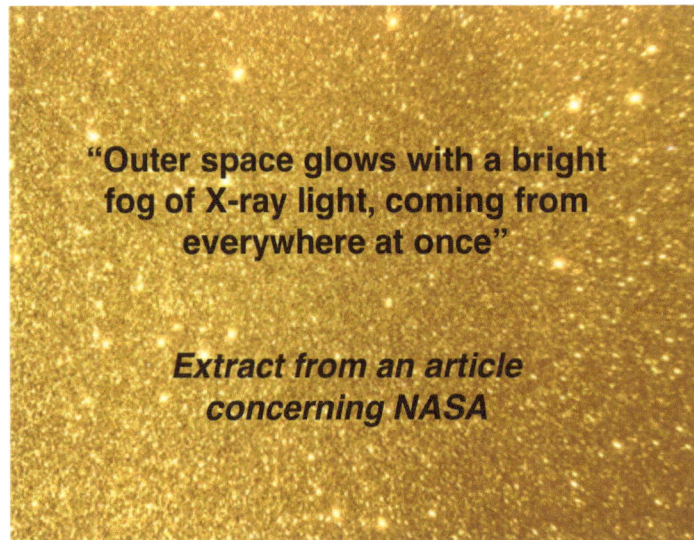

Fig 12. Outer space X-rays.

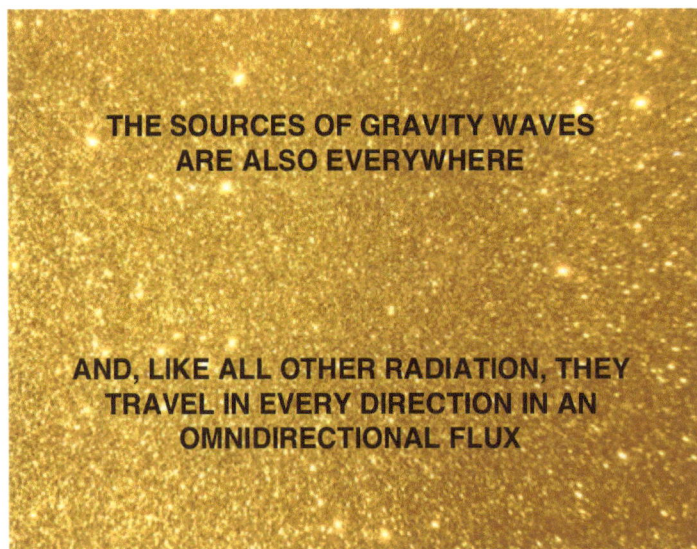

Fig 13. Sources of Gravitational waves.

2.3 TENET 3. That gravitational waves penetrate and pass through matter with great ease.

The third tenet is axiomatic. We have practical experience in our lives that, for example, several different wavelengths of EMF waves penetrate matter easily. Light penetrates water and glass, X-rays penetrate human flesh and bone, and so on. We know that gamma rays are even more penetrating, and are produced by stars, including our sun (see "Gamma Rays" in Wikipedia). There is no reason to expect gravitational waves to behave any differently. There is nothing scientifically prohibitive about a proposition that gravitational waves penetrate matter with consummate ease.

As noted above, the construction and use of the LIGO apparatus has demonstrated that gravitational waves do enter into matter and can cause sufficient drag on the internal atoms of that matter to change the length of the measuring arms of the enormous apparatus. Any professional criticisms of the initial wave analysis techniques—made in the true spirit of scientific research—will, naturally, lead to improved data quality and production.

In further, and more convincing, confirmation, it is also well known and recorded that gravitational waves penetrate dense matter such as rock with ease. We only have to travel two thousand metres down a mine shaft and into mine workings for us to be aware that we do not feel any difference in our weight, and if we pick up a piece of rock and drop it, it will fall at an acceleration of approximately 9.81 m/sec^2. This experience is hard proof, factually observed by thousands of miners every day. Gravity is operating all the way to the centre of the Earth and the centre of every star.

Thus, hard evidence supports the author's third contention in his theory, that gravitational waves penetrate matter with sufficient ease that they can pass down into and through planets and stars as if they were transparent.

Figure 14 is a diagram reminding us that light arrives at any given planet from all directions. Much of it is absorbed at the planet's surface, but a lot is reflected off back into space. This creates a zone that is bright near the planet and which becomes weaker further away from it as the reflected light spreads out and dissipates. After all, the moon's reflected light is still pretty powerful across the considerable distance between us. We can walk out in the moonlight on a summer's night and see our way by it, or, lacking a moon, even by starlight.

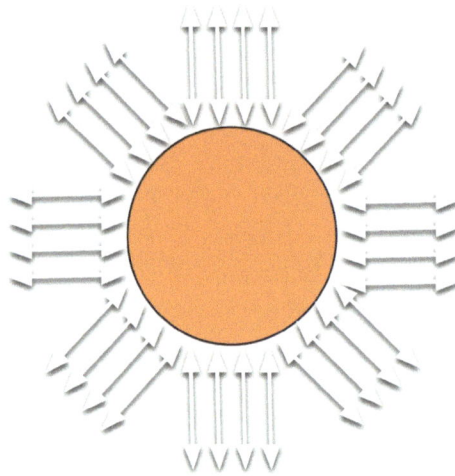

Fig 14. Light arrives from everywhere.

Figure 15 is a photograph reminding us that we can look up and see half of the Milky Way at any moment. That is 100 billion stars out of the 200 billion in our galaxy. And remember that there are quadrillions of galaxies out there with light arriving at Earth. Our naked eyes can't see them all, but our telescopes can. Wherever we are on Earth, we can look up and see the light arriving.

Fig 15. Light arrives from The Milky Way.

And this reminds us that gravitational waves will also be arriving at the Earth from all over the universe. But when they arrive at our planet, they don't get absorbed rapidly, and they don't get reflected off. They just pass straight through. Figure 16 shows a diagrammatic representation of omnidirectional gravitational waves passing through a planet.

The new concept contained in this theory is that gravitational waves pass through from all directions, thus creating a balance of forces. A single body is, consequently, not driven to move in any particular direction—only if it enters into the same locality as another body. Otherwise, in isolation, the forces maintain it as a sphere.

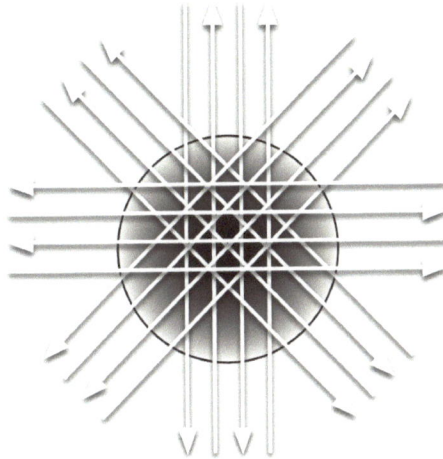

Fig 16. Gravitational waves passing through a planet.

Figure 17 is an artist's impression of an atom. In a typical very dense material, if the nucleus were drawn as being 1 metre in diameter, then the electron shell would be 10,000 metres in diameter. Everything in-between is pure vacuum.

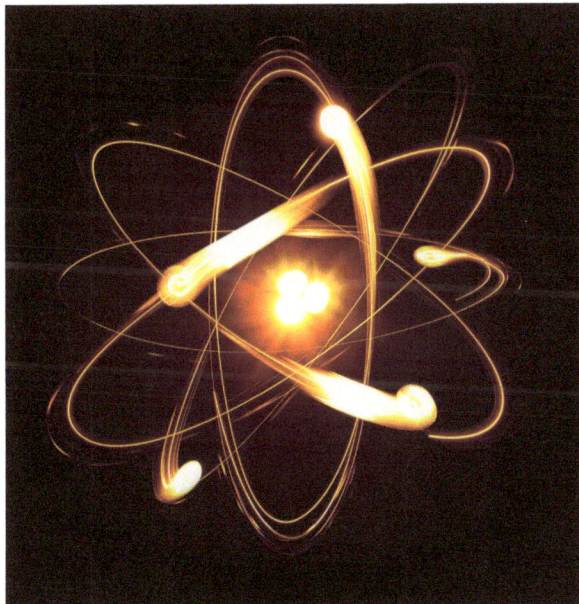

Fig 17. A cartoon atom.

The distance between the nuclei of closely-packed atoms would be 10,000 metres. That is, in one's imagination, a 1-metre ball positioned every 10 kilometres. Since everything between is vacuum, it is no wonder that gravitational waves can travel through matter. It is hardly different from interstellar space. And, of course, in Quantum Field Theory, the atom itself doesn't exist, being considered to be just a quantum field of its own, so that matter, as we know it, doesn't exist. The author has stayed with the concept of atoms for the sake of conservatism to be able to apply some meaningful scale to his proposals.

2.4　TENET 4. That in passing through matter, gravitational waves impose a tiny accelerating force on each atom in the direction of wave propagation, giving up a minute amount of energy in exchange.

There is factual evidence for electromagnetic (EMF) radiation being absorbed by matter as it passes through and interacts with it. Light is a useful analogy because we can see that its fundamental behaviour is indicative of how waveforms interact with matter.

We know that light is refracted and absorbed as it passes through water or glass, with a corresponding reduction in velocity. Light transiting through glass or water travels at a speed less than C, being the speed of light in a vacuum. And we know that light interacts with the atoms in water because it becomes gradually absorbed such that, at a depth of 1,000 metres below the surface of the ocean, it has become totally absorbed and there is no light left. Also, many of us have personal experience of being X-rayed and seeing the partial absorption of the X-rays on the negative images that the doctors examine. This demonstrates the interaction of EMF waves with the atoms of matter and their absorption proportional to the density of matter penetrated. These are facts. The author uses them to create the analogy with gravitational waves and shows how this has also been proven.

The LIGO experiments confirmed that gravity waves interact physically with matter and actually drag the atoms of the test apparatus so that they physically change length. See the following quotation taken directly, and unaltered, from the CALTECH LIGO laboratory web site, referenced below.

*"A gravitational wave does stretch and squeeze the wavelength of the light in the arms, but it turns out that doesn't matter. What matters is how long the laser beams spend traveling in each arm. When a gravitational wave passes, **it changes the lengths of the arms**, which changes how far each laser beam needs to travel before being reunited with its partner beam."*

*"The result of a misalignment in merging laser beam light is an interference pattern, and the level of misalignment tells us how much **the arms changed length** during the light-beams' journeys."*

SOURCE: https://www.ligo.caltech.edu/page/faq

The very act of LIGO reporting the existence of gravity waves is a direct recording of the penetration and interaction of the waves with the apparatus, altering its arm length, which change is measured. It is axiomatic that the gravity waves must interact with the metal of the arm because they have changed its length. The interaction is, necessarily, the creation of an accelerating force on each of the atoms of the material to produce the change of length that is measured by the lasers in the arms of the LIGO apparatus.

This physical interaction is the essential difference between gravitational waves and EMF waves. Gravitational waves cause a physical drag on the atoms of the LIGO apparatus, whereas EMF waves cause electromagnetic effects, but no drag. If light falls on a selenium bar, it changes its electrical resistivity, for example. That is an EMF effect.

The gravitational drag effect is not to be confused with the mechanical phenomenon observed in the Eisco Crookes Radiometer, as shown in Figure 18, in which light, falling on the vanes, appears to create a physical pressure causing them to rotate.

The radiometer consists of a partially evacuated glass bulb approximately 70 millimetres in diameter, containing a fine pivot which supports four light-weight metal arms. One side of each vane is blackened, the reverse side is bright.

Fig 18. Eisco Crookes Rotating Radiometer.

On exposure to sunlight or a sufficiently bright incandescent light source, the vanes turn. The absorption of radiant energy by the blackened sides of the vanes sets up a temperature differential with respect to the silver sides, causing a net force to be exerted by the Brownian movement of the air molecules on the blackened side which makes the vanes rotate with their bright faces leading.

The following is an extract from Roberts' 1978 monograph entitled, "A Theory of Gravitational Repulsion" (See Appendix 1).

"In a wave-form concept of gravitational repulsion, one must propound that, since the size of an atomic nucleus governs our current concept of "weight" [or mass], then gravitational absorption must be performed by the nucleus with very little effect - if any - from the electron shell."

Wikipedia says: *"Gravitational-wave observations complement observations in the electromagnetic spectrum. These* [gravitational] *waves also promise to yield information in ways not possible via detection and analysis of electromagnetic waves. Electromagnetic waves can be absorbed and re-radiated in ways that make extracting information about the source difficult.*

17

Gravitational waves, however, only interact weakly with matter, meaning that they are not scattered or absorbed [greatly]."

https://en.wikipedia.org/wiki/Gravitational-wave_astronomy#Development

It is the very weak interaction between gravity waves and matter that has delayed the ability of scientists to construct sufficiently sensitive apparatuses to measure even strong gravitational wave pulses. That interaction, which was speculated by Roberts in his 1978 monograph is now a proven fact. Thanks to the LIGO experiments.

This recent development, therefore, establishes the theory's proposed interaction of gravity waves with matter as a genuine, scientifically-recorded phenomenon.

In conclusion of the discussion concerning the tenets on which the theory is based, it has been demonstrated that they are scientifically simple, effective, and demonstrable processes that we see all about us in our scientific, engineering, and even domestic lives. In other words, observed facts!

When one considers that DOPA theory is based on these truths that lead inexorably to the differential opposing partial absorption mechanism, it becomes impossible not to recognise that Newton and Einstein's proposals were not mechanisms, only allegorical descriptions used to explain their amazingly prescient mathematical works describing how matter behaves in response to the force of gravity.

On the other hand, the proposals laid out in this book comprise a fully-working mechanism put forward for the first time by anyone (to the best of the author's knowledge) explaining how gravitational force (gravity) is inevitably produced. The word 'inevitable' is used because, since the four cornerstones of the theory are true (as has been demonstrated above) the behaviour of gravitational waves will inevitably produce gravitational force, as explained subsequently in this book. Absorption theory will also, inevitably, produce the phenomena of inertia and momentum, as propounded by the author in relation to the movement of matter through gravitational waves.

It is hoped, on this basis, that the reader will give serious consideration to this theory because, if accepted, it necessitates a rejection of the concept of spacetime and provides an alternative to the mechanism of spacetime distortion to explain the formation of gravity zones around material bodies such as planet and stars. There is no requirement that any changes should be made to the concepts of relativity for absorption theory to be valid.

CHAPTER 3 - HOW IS THE FORCE OF GRAVITY CREATED AND IMPLEMENTED?

3.1   Introduction to the omnidirectional gravitational wave flux—the universal gravitational field—and how it works.

You might think that the gravitational waves that travel everywhere form the 'gravitational field', but that would not be right because gravitational waves are not 'gravity', they are just waves that *create* gravity within matter. So, what is the field? You ask.

It is a field of 'POTENTIAL' gravity. You will see this word used a lot in this section of the book.

To try and explain it simply, if at a certain point, one gravity wave passing through a point in space is very strong, and the opposing wave passing through that point is equally strong, then they will balance themselves out. Believe it or not, this is actually a piece of 'information'. The piece of 'information' is simply the properties of each of the opposing waves. How so?

The information that the waves carry is only revealed and executed if they meet where, say, a molecule of matter is present. If the waves are equal in all directions, nothing will happen to that molecule in terms of being affected. The waves will each create a gravitational force within that atom, but each will balance out the other.

We could make it a little fancier and more complicated by creating a definition: *At any given point in the universe, the gravitational field's property is the vectored sum of all the gravitational wave strengths passing through it.* If they are all equal, then there is a 'null' effect at that point. And if a molecule of gas drifts through that point, then nothing will happen to it. It will just drift right through and go on its way.

Now consider a different scenario at another point in space where some strong gravitational waves are travelling in one direction and some weaker ones are travelling in an opposing direction. (The weaker waves may have passed through a planet a few thousand kilometres away and have thus lost some of their energy by being partially absorbed within that planet.) Now, the 'information' contained by the field at that point *is* that, if a molecule of gas drifts through that point, then the stronger waves will overpower the weaker opposing waves and will create a net drag on the molecule in their direction of propagation. Which will *automatically* be towards the source of the weaker gravitational waves. It is the depleted waves that create the potential imbalance, so when the potential is realised within the molecule, the vector of the force will automatically be towards the source of the weaker waves.

The waves thus hold, within them, the following information: a) what vector direction the net drag force will exert and, b) how powerful that drag will be. The net vector will depend upon the direction that the opposing waves are travelling, and the power of the drag will depend upon the difference in strength between the two opposing wave trains. This is a valid and true mechanism. No esoteric mathematical speculation is needed to understand this.

This information is there and only operates in the event that matter were to exist at that point. Thus, it is a *potential* effect. It is prepared and available to work *if* matter comes along, but does nothing otherwise. That is why Roberts calls it a potential gravitational field. The essence of this theory is contained in the following expanded definition:

"The universal gravitational field is a potential gravitational field composed of omnidirectional gravitational waves whose properties and vectors comprise the necessary information-in-waiting to create matching net drag vectors in matter that they penetrate. Those net drag vectors are the forces of gravity—gravitational force.

At any given point in the universe, the gravitational field's property is the vectored sum of all the gravitational wave strengths passing through it."

And that is it, in a nutshell. This is the simple explanation for how the universal gravitational field is created, what it is, and why Roberts calls it a *potential* field. Perhaps, now, you can see that the potential is present at all points, but is not an actual force until it is generated within matter.

So simple and elegant, without a single element of geometry or calculus needed to understand it.

And, in order to compare this theory with those of Newton and Einstein, you need to study what they can say about how information is transmitted throughout the gravitational field in their case. Why, you might ask, is it important to know this?

Because, somehow planets and stars have to 'know' where to go within the gravitational field when it affects them. Neither Newton nor Einstein could find the practical structure of the field, nor how it transmits the vital information. Roberts' theory, as explained above, can. With startling simplicity.

And, this elegant structure explains how matter can react to gravitational waves without having to send information at light speed (or faster-than-light speed); the potential gravitational field is already waiting in space for matter to come along and be affected. For example, the Earth moves around the Sun within a pre-formed and waiting potential gravitational spherical zone around the Sun that was formed eight minutes before the Earth moves into it, by weakened gravitational waves passing through the Sun and reaching the Earth's orbit. These waves are being refreshed continuously, and are being opposed by stronger incoming waves, leading to the Earth receiving a vector drag force towards the Sun.

This is the most exciting of theories because it defines how matter is controlled. It provides the answer for which we have been searching since before Newton's time.

The reader will be able to study a key example of why the transmission of information is a key factor in establishing the validity of any proposed theory of gravitation. See the last 19 paragraphs of Section 4.5. These paragraphs discuss the simple question of what would happen to the Earth if the Sun were to suddenly disappear. Only DOPA theory can answer this question with real, practical reasons for what would happen, and why.

However, in order to learn about all the different aspects of how this potential gravitational field creates planets and stars, and how it makes cosmic bodies try to move towards each other, you will need to read Chapter 3.

After that, you will need to fill out the picture and come to a full understanding of the amazing consequences of this concept; you will need to read right to the end of the book. There is critical information contained in each chapter right up to the last page.

So, in summary: Unlike the Newtonian and Einsteinian concepts, absorption theory provides a practical mechanism by which information is carried within the gravitational field *to create its effect*. It is the universally-present, omnidirectional flux of gravitational waves that comprises a field of *potential* gravitation. The field automatically adjusts to changes in the state of matter within it, and adjusts at the speed of light (because gravitational waves travel at the speed of light), but the information does not *connect* any two bodies, and therefore there is no connection or force between any two bodies causing them to behave in certain ways in relation to each other; *each reacts according to the pre-existing field state in its immediate vicinity.*

When no matter is present, gravitational waves do not interact with one another. That is just the same as omnidirectional light waves, which do not interfere with one another as they traverse the universe. It is only when 'matter' is present within the gravitational field that different-strength, pre-existing, gravitational waves penetrate 'the matter', pass through it, and impose an acceleration on its constituent atoms. That is when gravitational force is created. Gravity is neither an attraction between bodies, nor a repulsion between bodies, nor is it a pressure acting on the surface of bodies.

3.2 The speed of gravitational communication

If, within the wave flux, matter moves, is re-distributed, or is destroyed, as in a nuclear explosion, then the changes in the potential field differentials travel away at the speed of light as the 'message' is carried on outward-flying gravitational waves that migrate from the incident in the form of a sphere containing fresh information.

The 'information' that controls matter is the difference, at any point within the gravity field, between the relative strengths of all omnidirectional waves passing through that point. If any one or more of those waves has a smaller amplitude than any of the others, then there is not a full cancelling-out of potential gravitational force and, therefore, a potential force vector exists that will affect and control any matter that finds its way to that point. This wave amplitude difference is the information that forms the gravity zones within the near-infinite gravitational field that controls cosmic matter throughout space.

The mechanism for the creation of the gravitational field and its information transmission system is one of the prime contributions of Roberts' theory to science, and which contributes an alternative view of spacetime's gravity well concept. This knowledge allows us to contribute to the basis for the theories of relativity while enabling us to continue our lives in a three-dimensional world where time is something we can just watch go by as we get older.

The fundamental development that this book represents is that it describes real, practical, mechanisms by which gravitational information is created, coded, and transmitted outward from any one part of space to another at the speed of gravitational waves, Cg.

Thus, gravitational force does not travel at any speed at all—it simply exerts itself because *gravity is the acceleration response created in matter when gravitational waves pass through it. Gravity does not exist outside of matter; it is not a force linking objects.*

There is no direct, mystical, 'connection' between the atoms of the sun and the atoms of the Earth, as proposed mistakenly by Isaac Newton. It is the sudden cessation of the partial absorption of gravitational waves by the Sun that the Earth would feel 8 minutes later if the Sun were to disappear. (This is covered in the last 19 paragraphs of Section 4.5.)

The gravitational 'shadow' between the Earth and the Sun would simply travel past the Earth, and the Earth would, at that time, cease to be held in orbit and would be released to travel tangentially away from its circular motion. At that same instant, the last of the EMF radiation from the Sun would also fly past the Earth, and the gravitational and light effects would spread simultaneously through the Solar System and outward at the speed of light C and gravitational waves Cg.

3.3 Background to the existence of gravitational waves

The discussion is simple: the universe is full of gravitational waves. These were probably created in the big bang, and are still there in the same way as is the Cosmic Background Radiation that scientists measure today. It is presently considered that gravitational waves can also be created by large accelerations or changes in matter. Alternatively, it may, ultimately, be found that gravitational waves are emitted by matter itself.

However, it is important to note that the origin of gravitational waves does not affect the construction or outcome of the author's theory.

Roberts' paper of 1978 postulated that gravitational waves existed and that they cause the phenomenon of gravitational force. He did not speculate then on their origins and, even now, he comments only very tentatively on the various possibilities.

3.4 The proposed mechanism for the creation of gravitational force

This theory does not clash with Einstein's theories of relativity; it newly contributes the one thing that he could not explain—the concept of a working mechanism for the creation of gravitational force which is described in relativity as a distortion of spacetime by matter.

In 1978, Roberts thought that gravitation could be a repulsion, but has subsequently changed that concept to an internal 'drag' effect. A more detailed and informed consideration of the possibilities showed that gravitational waves cannot function by simply creating a surface 'push' on any matter that they encounter. There is just no acceptable engineering mechanism by which this might be effected and, furthermore, the effects of such a mechanism for movement of matter would not match those observed within, for example, engineering structures or within planets.

In the greater-central part of the universe, absorption theory proposes that the gravitational wave flux is omnidirectional, homogeneous, and steady, as justified earlier. That means that waves are travelling in all directions uniformly at a consistent density. That is not surprising, considering the 13 billion years that they have had to mix and travel. This premise conforms with what we observe around us; for example, planets and stars are formed spherically, indicating uniform (omnidirectional) forces to be employed in their formation and permanent state.

The author's theory is that gravitational waves enter into and pass through matter—for example, the Sun, the Earth, you, a cannonball, or a feather—interacting with the constituent atoms and creating, internally, an acceleration on those atoms in the direction of wave propagation.

Both the lack of a coherent theory and the amazing penetration capability of gravitational waves into matter have, heretofore, made it impossible to observe them directly, or to build equipment to do so. Everything is exceedingly transparent to them, so trying to study them was, until 2016, impossible in terms of direct observation. It is still possible that what the LIGO apparatus is currently measuring is a disturbance modulation imposed on, and travelling through, the gravitational wave field in much the same way as we use modulation in FM radio to transmit a signal through a very much shorter wavelength radio wave carrier.

Figure 19 is a figure from the author's 1978 paper, being a graphic image of a gravitational wave interacting with a constituent atom of matter and consequently losing a minuscule fraction of its energy. As the wave train continues to pass, it continues to exert a tiny drag (accelerating force) on the nucleus of the atom. Since every constituent atom is accelerated, the entire body is accelerated simultaneously and uniformly—even if it is the size of the Earth or the Sun.

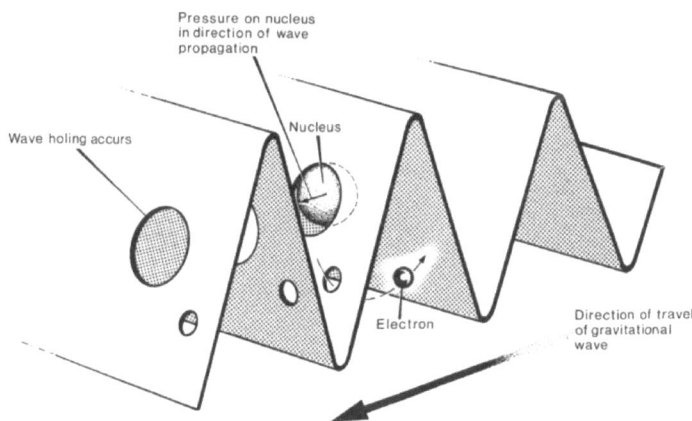

Fig 19. Wave holing, from Roberts' 1978 paper.

In recognition of Quantum Field Theory, Figure 20 illustrates an alternative image of multi-frequency waves imposing drag on quantum field packets, which replace atoms. It is likely to be easier for qualified physicists to find a detailed mechanism for gravitational waves to affect quantum field packets than to affect the physical 'matter' of atoms.

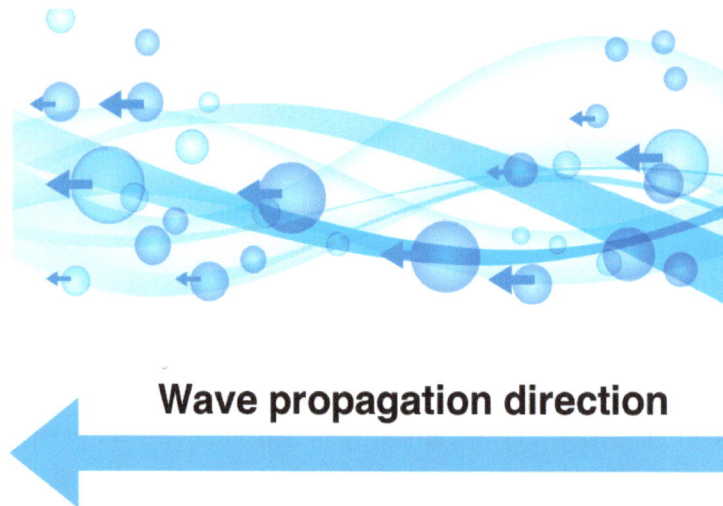

Wave propagation direction

Fig 20. Wave propagation within matter, dragging discrete quantum fields.

This concept of imposing acceleration onto individual atoms at the ultra-microscopic level permits gravitational force to affect individual atoms, molecules, or specks of dust in space. It also allows the application of gravitational force to gasses such as our atmosphere and liquids such as our oceans. Any concept of surface pressure would not work with them, and so would not be feasible.

Einstein could not provide any mechanism to account for the presence of gravity wells around planets and stars—which we all recognise exist—so he adopted the idea of 'spacetime' from Minkowski's paper of 1908, who, himself, had copied the idea from Poincaré who published his original concept in 1905/6. Having co-opted the idea from these two, Einstein imagined that matter creates a gravity well in spacetime around itself, wherever it is. That was the same mistake that Newton had made in trying to account for the now-discredited, mysterious, attraction of all atoms to all other atoms in the universe. 'Spacetime' did not exist before 1905 and was not 'discovered' in the same way that Professor Henri Becquerel discovered X-rays (and which concept was developed by Pierre and Marie Curie), which did exist prior to its discovery, or in the same way that Livingstone discovered (as far as Europeans were concerned) the magnificent waterfalls in Africa that he named the Victoria Falls. In other words, spacetime is a science-fiction invention that came true. Since it led to the wonderful theories of relativity, no one cares much to look at it too closely—particularly as to how we can link it to the real world.

The fact remains that Einstein adopted something which had no reason to exist and without any mechanism by which it operated—spacetime and its distortions. An imaginary 4-dimensional cosmos with time as the 4th dimension. The entire idea was an imaginary construct supported by mathematical equations that work beautifully and have been tested to work, so the only problem is the questionable premise on which they are based.

Figure 21 shows the concept of a gravity well around a planet. It is an imperfect analogue because the 'well' is a three-dimensional representation of the distortion of four-dimensional spacetime around a planet. *(The author hopes that the term 'gravity well' might, in future, be replaced by the term 'gravity zone'. In diagrams this should be shown as a 3-dimensional sphere with rapidly-decreasing intensity away from the object/planet/star being considered.)*

24

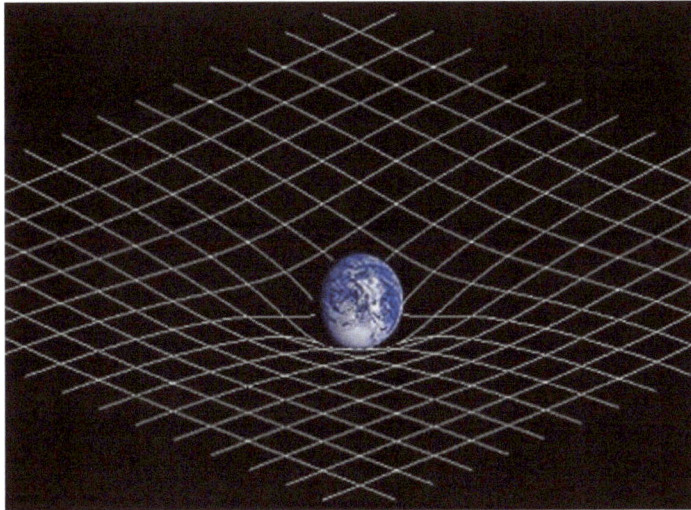

Fig 21. An artist's impression of a gravity well around a planet.

This type of '2-dimensional mesh' artistic illustration is always drawn incorrectly in that it shows the maximum gravitational force to be at the centre of the planet, whereas Newton demonstrated, and it is still currently-accepted science that the net gravitational force is at a maximum at the planet's surface or within the planet, falling to zero at the centre. The author stresses to beginners who see these diagrams to recognise that they are usually in error in more than one way and beginners should not take them literally as having any scientific value.

The problem, under spacetime theory, is not in recognising that gravity wells exist, nor in imagining them, but in explaining what has caused them to exist, and what tries to make matter move towards other matter down that gravity well. What is the force? And how is it created? Neither Newton nor Einstein could provide a reasoned mechanism to answer these two questions. The author proposes that such a lack makes both offerings unacceptable and offers DOPA theory as containing both a reason for its existence and the essential working mechanism.

In DOPA theory, in the case of a single planet, for example, the constant absorption of waves from every direction accelerates the constituent matter of the planet towards its centre and holds it there under compression. That is how the planet formed in the first place. Similarly, any object or person standing at the planet's surface is trying to accelerate itself down towards the centre of the planet by the drag force being exerted on its atoms by the incident, transient, gravitational waves. The difference between waves coming in from outer space incident onto the planet's surface and the waves that have passed through the planet and are coming, partially-absorbed, outward, is the net force (attempted acceleration) of gravity experienced at the surface of the planet—in our case, being about 9.81 m/sec^2.

Let us look at it more personally. You are standing on the surface in the U.K. with full-strength gravitational waves coming down through your head and trying to drag you down onto the Earth's surface at 100% pristine strength. At the same time, gravitational waves that enter the Earth in the antipodal position of New Zealand pass through the planet, lose some of their energy, and exit into you from beneath your feet with an attempted, induced, upward acceleration of, say (fictitiously,

25

for the sake of argument) 99% of pristine. The waves that come down through you are stronger than the waves that come up through you, so there is a net difference in force of 1% between the two that results in a net drag downward. We call this gravity. It is straightforward to understand. In the case of the Earth, the factual difference is 9.81 m/sec^2, but we have no idea, at present, what the actual numerical value of the 100% pristine wave strength is. That is a subject for necessary research.

The author proposes that *potential* gravity zones actually exist and thus do not have to be created by a theoretical concept. He does not argue with or dispute relativity; instead, he provides an acceptable physical reason for gravity zones existing and a reasoned mechanism for how they are formed and maintained in terms of information transmission. By describing these mechanisms, the author supports the concept of spacetime by providing a new foundation for it.

At this point, it is necessary to examine those mechanisms in detail.

Figure 22 shows a single, isolated, body in space and considers a single gravitational wave (bold red arrow) passing through the planet from left to right (absorption graph above).

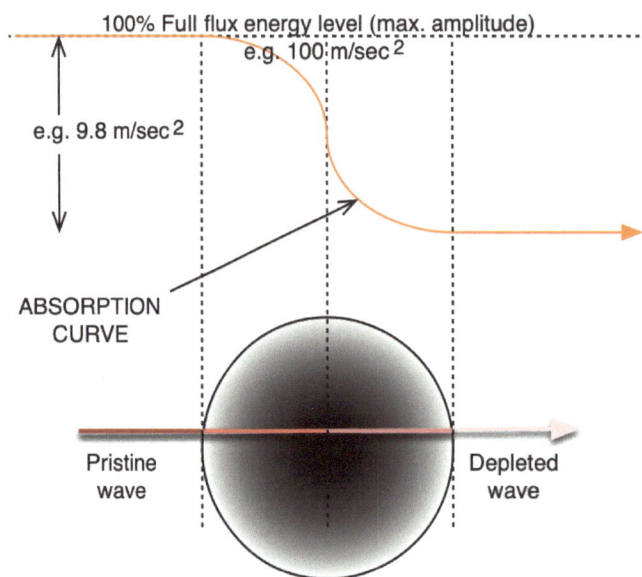

Fig 22. Partial absorption of a gravitational wave.

If this were the planet Earth, then this gravity wave would strike the surface of the Earth at a 100% energy level, i.e. maximum amplitude which, for the sake of argument, we could call 100 m/sec^2 *potential* acceleration at the point of strike, being a fictitious figure.

Passing into the planet, as shown, the gravitational wave would start to be absorbed by travelling through first the outer zones of the planet and then the denser outer and inner core, being increasingly absorbed by the increasingly dense nature of the Earth's interior until the wave reaches the centre of the planet. Reference to the graph at the top of Figure 22 shows that having passed the centre point of the planet wave absorption continues through the core, decreasing towards the surface of the planet. The wave loses energy all the while, until it emerges from the planet with a reduced amplitude equivalent to 9.8 m/sec^2 and thus having an outgoing force of, say, 90.2 m/sec^2.

If this were the only wave to be considered, then the planet would be ruptured towards the right, and any object on its surface would be accelerated away at 90.2 m/sec^2.

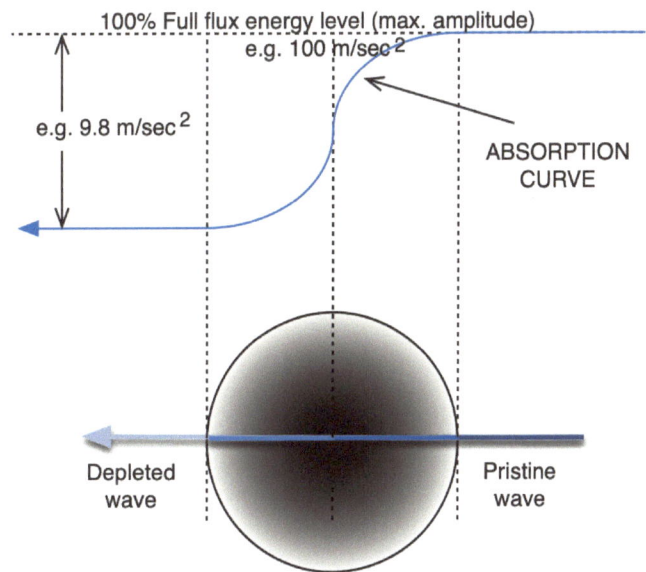

Fig 23. Partial absorption of an opposing gravitational wave.

Fortunately, that is not the only wave to be considered. Because the gravitational wave flux is omnidirectional, there is always an equal incoming wave travelling in the opposite direction, as illustrated in Figure 23 (coloured blue). In this case, the wave travels through the planet from right to left, creating an opposing acceleration in the planet's constituent atoms. (This wave's absorption graph is also shown above.)

Figure 24 shows the gravitational outcome of the mutually-opposing incident gravitational waves.

In that figure, it can be seen that each opposing wave enters with an amplitude equivalent to an imaginary 100 m/sec^2 and each leaves with an outgoing amplitude of 90.2 m/sec^2. These opposing forces lead to a net balance of an inward acceleration of 9.8 m/sec^2, which is approximately the acceleration of gravity that we experience at the surface of the Earth.

It is to be noted that we do not presently know what the full 100% flux force represents in terms of potential acceleration, but it is immaterial to the principle of the theory because what is essential is that there is a differential opposing partial absorption that results in a net balance of 9.8 m/sec^2 downward in our case, as detailed in Figure 24.

27

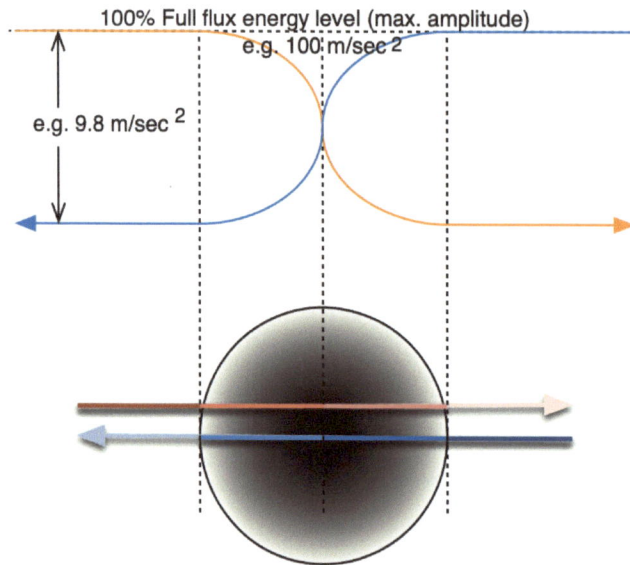

Fig 24. Balanced and opposing gravitational waves.

It would not be impossible that the incoming potential acceleration was 1,000 m/sec², because it can be observed that there will always be an opposing 1,000 m/sec² wave, so it is the absorption value that is the critical feature that results in a net inwards acceleration of 9.8 m/sec² in our case.

In the case of our moon, since the absorption is less, the surface acceleration of gravity is less. In Figure 24, representing Earth, a field of 100 units is reduced by absorption to 90.2 units. In the case of a less-dense, smaller-mass planet, the incoming field of 100 units would be absorbed less by the less-dense, smaller mass, so the outgoing wave is degraded to only, say, 98 m/sec². The net difference between the incoming and outgoing wave force acceleration is therefore only 2 m/sec² inwards. Thus, gravity zones around different mass planets have different, proportional, maxima depending on the planets' mass, density and radii. In the case of our moon, this results in a net inwards surface acceleration of 1.62 m/sec².

Notice in Figures 22 to 24 how the steepness of the curves indicate that the rate of absorption is related to the density of the material through which they are passing, and not just on the total amount of mass encountered, meaning that a small body having the same total mass as a larger body will exhibit a greater gravitational force (and hence imposed acceleration) at its surface than the larger body. This is in accordance with Newtonian principles.

*This simple concept of differential opposing partial absorption is the first key concept of this new theory. It is **very** important because, heretofore, no one has been able to provide a mechanism whereby a planet controls its own creation of gravitational force. This theory now says **why** planets of different mass have different gravitational fields that peak to a maximum at the surface and **why** that surface maximum differs from planet to planet or star to star. Note that it is not just the mass, but the density of the mass that controls the gravitational wave absorption rate, as shown in Figures 22 to 24. Thus, the same mass within a decreasing radius object will demonstrate an increasing gravitational surface acceleration through an increasing force F, matching the*

28

Newtonian equation of $F = G\,M_1\,M_2/r^2$. However, some interesting inadequacies of this equation are discussed in Section 5.7 of this book.

3.5   A new explanation of shell theory that replaces the Newtonian inverse square law interpretation.

The currently-accepted 'shell theorem', developed by Newton, stated:

"If the body is a spherically-symmetric shell (i.e. a hollow ball), no net gravitational force is exerted by the shell on any object inside, regardless of the object's location within the shell. [This should have read, "within the shell's hollow interior".]

Newton, necessarily, considers that the shell exerts a gravitational force on itself. This follows from his theory of gravitational attraction. This, alone, negates the validity and viability of Newton's theory. It is recognised that Newtonian attraction does not work and that it has been long-ago replaced. Therefore, by simple consideration, we should not be accepting it and quoting it today. The problem that we have is that there is nothing adequate to replace it, and so it is used in a 'Nelsonian' way, by turning a blind eye to the contradiction. Like relativity, it works, so don't question it.

Beyond that complaint, there is the contradiction inherent in Newton's attraction principle when exercised within a planet. That is that when at the centre of a solid planet, the net gravitational force is outward in all directions. It cannot be any other way. From the centre, all mass is situated outward, and therefore, it cannot be any other than that all attraction force is exercised from outside the central position. That is a paradox. How can there be any pressure at the centre of the planet if all force at the central cubic metre is pulling it outward? It is the weakness of attraction theory.

Figure 24 shows why absorption theory produces a plausible nil net gravitational force at the centre of the planet, without the failing of producing outward-vectored forces. That figure shows, much more sensibly, that in DOPA absorption theory all imposed net gravitational forces within the planet remain inward vectored, which is theoretically sound and sensibly logical, including the reason for a zero vectored force at the planetary centre.

In this regard alone, absorption theory is a major improvement on past ones.

Similarly, the spacetime concept falls down when considering forces within a planetary mass owing to the lack of adequate or any explanation for how matter is supposed to follow distorted spacetime within matter when, at the same time, the presence of that matter is distorting matter. It is too convoluted to be acceptable.

Then there is the fact that Newton says nothing in general terms about what is going on within the outer shell being considered in his shell theory. Naturally, Newton used his newly-invented calculus to perform the calculation to prove his theory. This platitudinous reasoning is the thing that everyone repeats today about relativity. They say that because the mathematical calculations prove the idea to be correct, then it must be correct. And yet, we are obliged to recognise that (like relativity) if the input data are wrong and invalid, then the output data are, similarly, wrong and invalid. And, the output conclusions are wrong and invalid.

What did Newton do that was wrong? He used attraction as his force to perform his calculations. Since attraction does not exist, his conclusions are not valid.

Similarly, relativity contains many contradictions including 'time dilation'. This part of relativity says that time itself must run faster wherever there is a relatively decreased net gravitational force and slower wherever it is relatively increased. So, what happens when that force starts to fall off towards the core of a planet? At that point, owing to the lack of net force, time should be running (infinitely?) fast. No matter how one views it, spacetime means that time varies within the confines of a planet or star. Such abstruse and unlikely concepts reflect poorly in contrast with the simplicity and rational nature of DOPA theory which does not produce such conflicts and contradictions.

Secondly, examination of Figure 24 above shows that, in DOPA theory, net gravitational force continues to be exerted inwards all the way to the core centre, where it becomes zero by virtue of matching internally-directed absorption. At the centre, the graphs show that there will automatically be zero gravitational force whilst allowing a correct explanation as to why pressure increases within the body towards its centre.

Absorption theory explains why at any point going down towards the planetary centre in either a homogeneous, graded, or densely-cored sphere, the vector inwards gravitational force reduces. This conflicts with Newtonian theory as applied using Newton's famous (but now discredited) equation. Inappropriately using that equation, the PREM curve shown in Figure 25 proposes an *increase* in net gravitational acceleration with depth below the Earth's surface as deep as the outer surface of the outer core. Absorption theory, on the contrary, explains the existence of a *reducing* gravitational force downward with the additional benefit of maintaining inward-vectored gravitational force from the homogeneous planet's surface to its core. Moreover, absorption theory is the only one that provides a realistic working mechanism for the observations of gravitational variation *within* a planet or star.

The adoption of DOPA theory disagrees with the current Preliminary Reference Earth Model in which, unintuitively, the net gravitational 'pull' of the Earth *increases* downward at first because a dense core exists—as illustrated in Figure 25 below. Figure 25 is correct as a forecast only if one accepts the use of Newton's inverse distance theory to calculate the 'residual attraction' of any central sphere after discounting its shell. Conversely, a continuously *decreasing* net gravitational force inwards is expected towards the core using absorption theory, based on his explanation of differential absorption of gravitational waves to provide the working mechanism. *Put simply, Newton's inverse square law of attraction is not valid within a planet, and should not be applied to produce the PREM.*

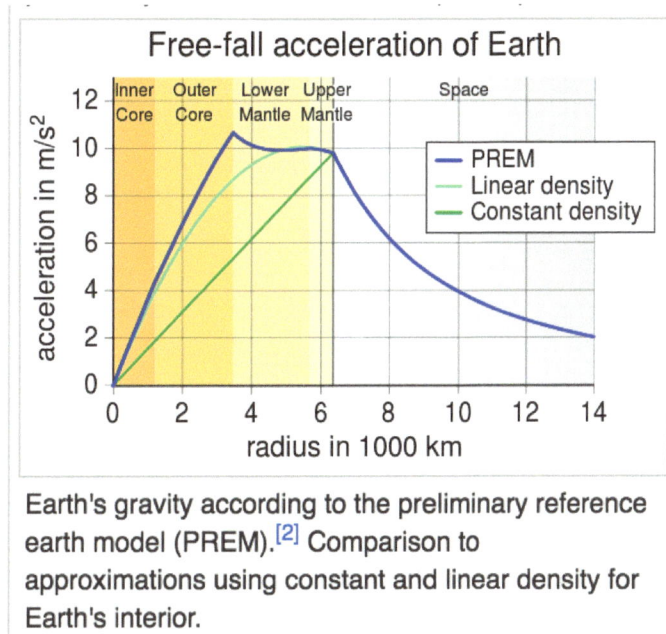

Free-fall acceleration of Earth

Earth's gravity according to the preliminary reference earth model (PREM).[2] Comparison to approximations using constant and linear density for Earth's interior.

Fig 25. Preliminary Reference Earth Model based on Newton's equation and law of attraction.

SOURCE: https://en.wikipedia.org/wiki/Gravity_of_Earth

Derivation of Roberts' shell theorem.

As demonstrated in Figure 26, the author's Shell Theorem adequately parallels Newton's Shell Theorem in that it accounts for the lack of gravitational attraction at any point within a hollow, spherical, shell body. It is just that the mathematics are fundamentally different, being based on a sensible working mechanism—which is of importance even though the results are identical. DOPA theory produces inward balancing forces from the surface to the core, but Newton's shell theory necessitates outward balancing forces. It is proposed that Roberts' Shell Theorem can be successfully used to replace Newton's.

Note how, in the 'shell' part of Figure 26, equal-and-opposite penetrating waves are partially absorbed in the shell—as if it were a solid sphere—and gravitational acceleration towards the hollow core centre (measured within the shell material) continues downward as absorption progresses.

Then, entering into the empty void, the waves pass directly across that void with constant reduced values of amplitude and energy. All waves balance one another out in that void, leading to zero potential gravitational force.

Upon reaching the opposite side of the void, the waves enter the shell and continue to be partially absorbed on their way out, exiting at the outer shell surface.

Figure 26 also shows why, under this theory, the overall outer surface gravitational force of the shell planet should be less than if it were solid. It has a reduced mass, and therefore a reduced gravity at its outer surface. This theory does not contradict the principle of 'shell theory', it explains

31

why it is valid, and it avoids having to resort to Newton's concept of *gravitational attraction* to explain it. Conversely, absorption theory explains it perfectly and soundly.

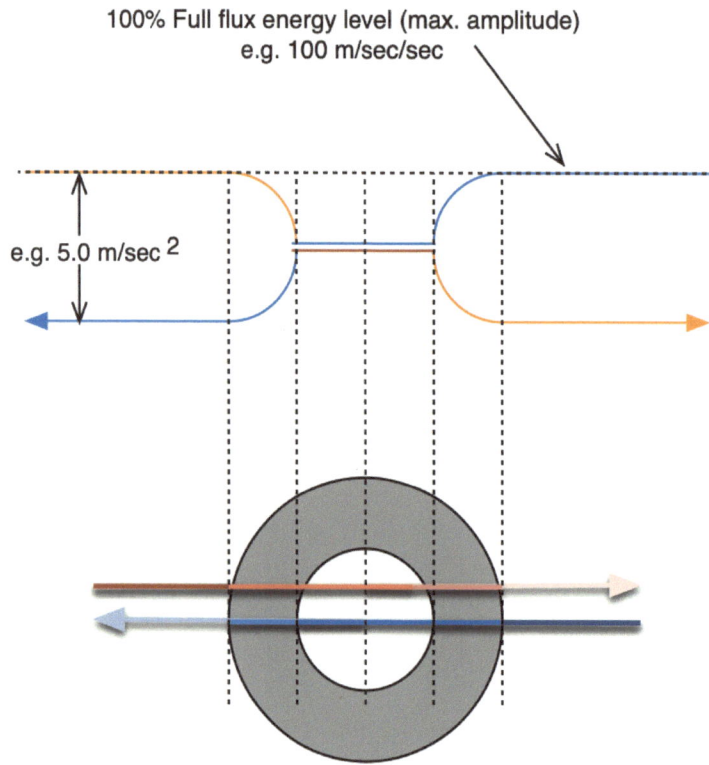

Fig 26. Explanation for the validity of absorption Shell Theorem.

What is significant is that absorption theory replaces the need for the concept of the 'attraction of matter-to-matter', as proposed by Newton. This new differential absorption theory explains shell theory perfectly without having to resort to a mysterious, inexplicable, attraction force between atoms across the universe. It also avoids the contradiction inherent in the attraction theory that, at the centre of a planet, all gravitational forces must be directed outward radially, not inwards because all of the mass is distributed spherically around the observer.

3.6   The potential-gravitational field forms gravity zones.

The concept of differential gravitational wave absorption provides a mechanism that meets all known observed phenomena including the formation of external potential gravity zones (currently called gravity wells) above the surface of, and extending out spherically from, cosmic bodies with a decreasing potential gravitational force approximating to an inverse square law.

The formation of potential gravity zones is one aspect of absorption theory that might, one day, be subject to experimental verification. How that could be done is not imaginable with our present state of technology.

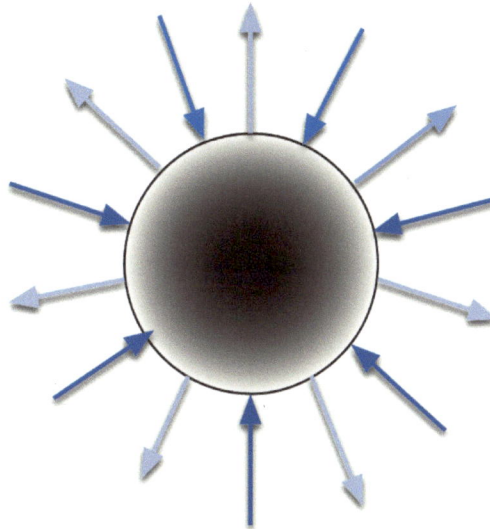

Fig 27. Artistic impression of wave trains entering into and exiting from a planet such as Earth.

Figure 27 is an attempt to show that the environment around a cosmic body comprises a three-dimensional flux of full-strength waves entering the body (dark blue), and a three-dimensional flux of degraded, partially-absorbed, waves leaving the body (light blue).

In absorption theory, degraded waves are continually emanating from the body at the speed of light, forming a sphere of potential gravitational imbalance around the body. This sphere will interact with other such spheres to create the situation shown artistically in Figure 28 below.

In the case of the Earth, absorption theory explains why the surface gravitational acceleration reduces from approximately 9.81 m/sec² with distance away from the surface as is known to happen, and as is shown in Figure 25. The reason is explained in this theory by the outgoing degraded waves becoming dissipated and 'diluted' in an increasingly-large sphere expanding outward. This sphere, referred to as the external potential gravity shell, does not exist in the form of a traditional 'field' because it consists of non-interfering gravitational waves of differential strengths that do not generate a gravitational force until they enter into a body consisting of matter. Whether that matter is comprised of atoms or quantum fields is immaterial to the outcome of the theory.

The force within a body, once created, continues to exist within that body only as long as incoming waves penetrate and pass through it—which is 'always' in practical terms—as long as the matter concerned exists within the ubiquitous universal omnidirectional flux of gravitational waves. This permits a very similar inverse square law of distance to operate in absorption theory as in Newton and Einstein's concepts. All three theories parallel one another in this regard, but only absorption theory explains in simple terms how and why the gravitational force is generated and why the potential force reduction outward from the body takes place. Newton and Einstein simply said that it happens and provided mathematical proof for their statements, but no mechanisms.

What is extending out into space, and becoming diluted, is not an actual force of any kind, it is a 'potential differential force' that *would be applied* to any body of matter existing within it or entering into it. It is another critically important part of absorption theory that the concept of a pre-existing potential flux eliminates the problems experienced in accepting either Newton or Einstein's response-time problems. In absorption theory, a potential is established before matter enters that part of space, and so, appropriate internal acceleration is initiated instantly within any body of matter. There is no connection as such between different bodies. For Newton's theory to work, every atom in the universe must somehow, inexplicably, be attracted to every other atom across the entire universe and for this to operate, that attraction would have to operate instantly—infinitely faster than the speed of light. It is one more major stumbling block in Newton's work. In the case of Einstein, the spacetime distortion does not pre-exist, as do the gravity zones in absorption theory; spacetime distortion, apparently, spreads out at the speed of light but is created inexplicably and instantaneously by the 'very presence of matter'; a feat made possible by the invocation of time as a fourth dimension of reality. Einstein's spacetime spreads out in the same spherical manner as Newton's attractive force, but its explanation is a theoretical amalgamation of space and time to create a 'new reality'. Absorption theory is, in contrast, pragmatic, down-to-Earth, and works in a known three-dimensional reality within known physical processes that are explained in this book. The author, for the first time, offers a sensible alternative to that of advanced theory.

Upon reflection, considering the immense size of some astronomical bodies that are controlled by gravity, it seems likely that the ultimate value of the undegraded gravitational potential force of the wave flux is likely to be considerably more than the notional 100 m/sec^2 chosen in the above example illustrations. It must, eventually, be found that the undegraded value of gravitational waves is (for example) more like 10,000 m/sec^2, in which case the Earth-degraded outward force would be 9,990.2 m/sec^2. If this were not so, then there would be insufficient differential leeway to account for planets and stars that experience a very high net surface gravitational force. After all, the surface gravitational acceleration of our sun is about 275 m/sec^2, and there are much more massive bodies out there in space.

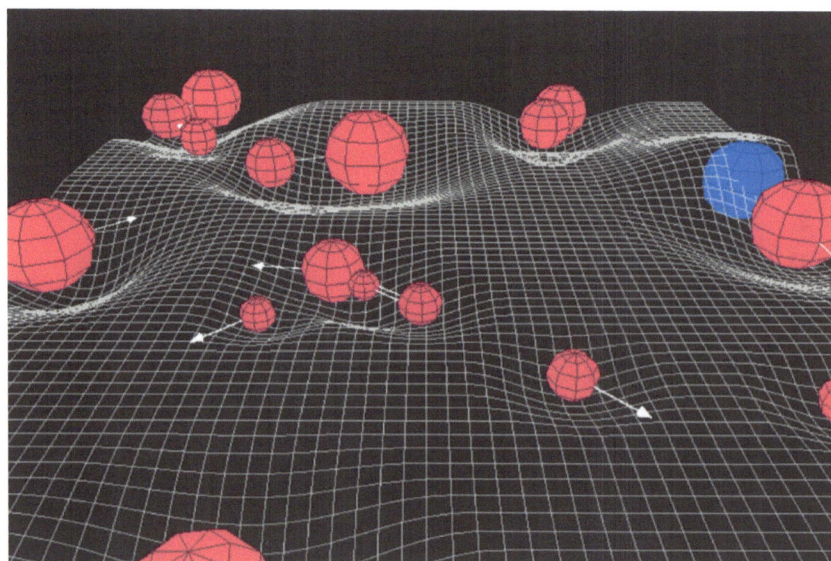

Fig 28. Artistic representation of multiple 'gravity wells' as a warped two-dimensional net field. (Credit Wikipedia)

Since each cosmic body is surrounded by an external potential gravity halo (currently, inappropriately, called a gravity well) of decreasing potential-force value with distance, then each must be able to interact and relate to others, such as in the case of our Sun, the Earth, our Moon, and the other bodies in the solar system. Artists try and illustrate this with the two-dimensional image of a distorted net, such as that in Figure 28. It is unrealistic, but it helps. It gives some assistance in demonstrating how potential-gravitational plateaux exist between cosmic bodies, where the gravitational gradient starts to dip in different directions towards each interacting body.

The author feels that the diagram in Figure 29 gives a better visual conceptualisation of an external potential gravity halo than does the two-dimensional mesh of Figures 21 and 28. It is likely, he feels, that the term 'halo will give rise to much less confusion among non-specialists than does the term 'well'. Public questions have been recorded such as, "What is at the bottom of a gravity well?" This is not useful in assisting understanding. Alternatively, the term 'zone' might be applied preferentially to 'well'.

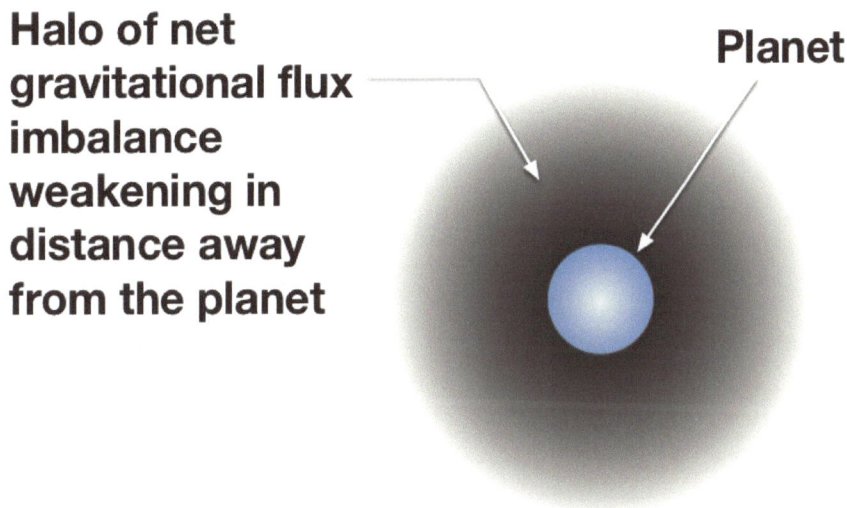

Halo of net gravitational flux imbalance weakening in distance away from the planet

Planet

Fig 29. Artistic representation of an external potential gravity zone.

The decreasing halo effect is not a force field per se. It is a zone of decreasing difference in incoming and outgoing gravity wave strengths (amplitudes). When an object such as a satellite is placed into that flux matrix, then it experiences a differential incoming-to-outgoing force within its own body, and the net value is the inwards gravitational centripetal force it experiences and which it has to counteract by its orbital velocity and consequent centrifugal force.

3.7   What makes planets and stars try to move towards one another?

What we have covered so far has explained how gravitational force is created deep within a planet, at the planet's surface, or in space away from the planet on a decreasing basis with distance away. We have also studied how gravitational force affects a single, isolated, body. The theory has explained why interstellar dust clouds coalesce to form stars and planets. And yet, why under DOPA theory, should any two cosmic bodies want to move towards each other?

In the case of any two neighbouring objects, such as planets, we only need to consider the waves that pass through both planets sequentially. These are the waves that have passed through one planet before entering the other in a diminished condition. All other waves are superfluous. Those that only pass through one planet still act to create internal gravitational stability to that planet as described above, but those waves that pass first through one planet and then through the other (as shown in Figures 30 and 31) are the ones that cause the planets to try to move towards each other.

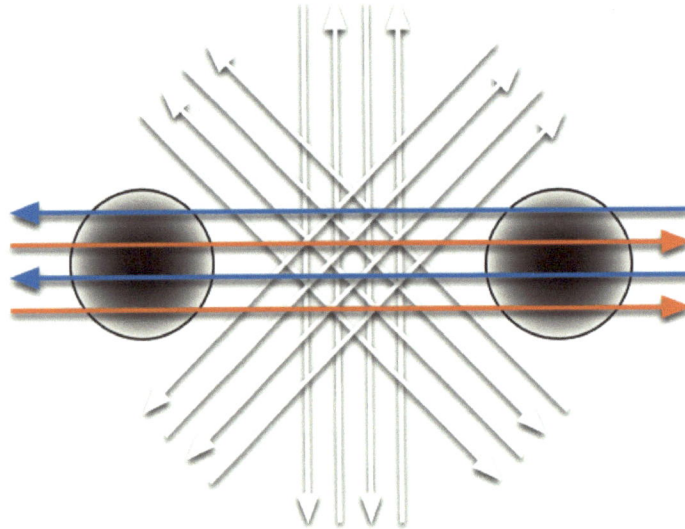

Fig 30. Artistic representation of two-planet absorption of gravitational waves.

In the case of a pair of planets (Figure 31) the intervening space between them contains only degraded, reduced-potential-force waves. When these enter either planet, they can only cause a weak acceleration away from each other (lighter blue & red), whereas the incoming acceleration from fresh, pristine, waves is greater (darker blue & red). The LHS planet has fresh red waves driving it to the right, but only weak blue waves driving it to the left. In the RHS planet, dark blue pristine waves drive it to the left, and only weak red waves try to drive it to the right.

Therefore, there is automatically a net balance of internal acceleration towards each other.

Each planet automatically tries to move towards the other by dragging itself into the gravitational 'shadow zone' between the two. There is no force or connection between the two, just a 'shadow' zone into which they must automatically, and independently, move.

This is a straightforward and logical mechanism. Neither planet is connected to the other; they move independently, unaware of the other planet in their vicinity. All that each planet experiences is the dearth of resistance from a particular direction towards which it, naturally, accelerates. In the case of multiple bodies, the net vector of halo interaction forces will result in the single desired direction of movement.

36

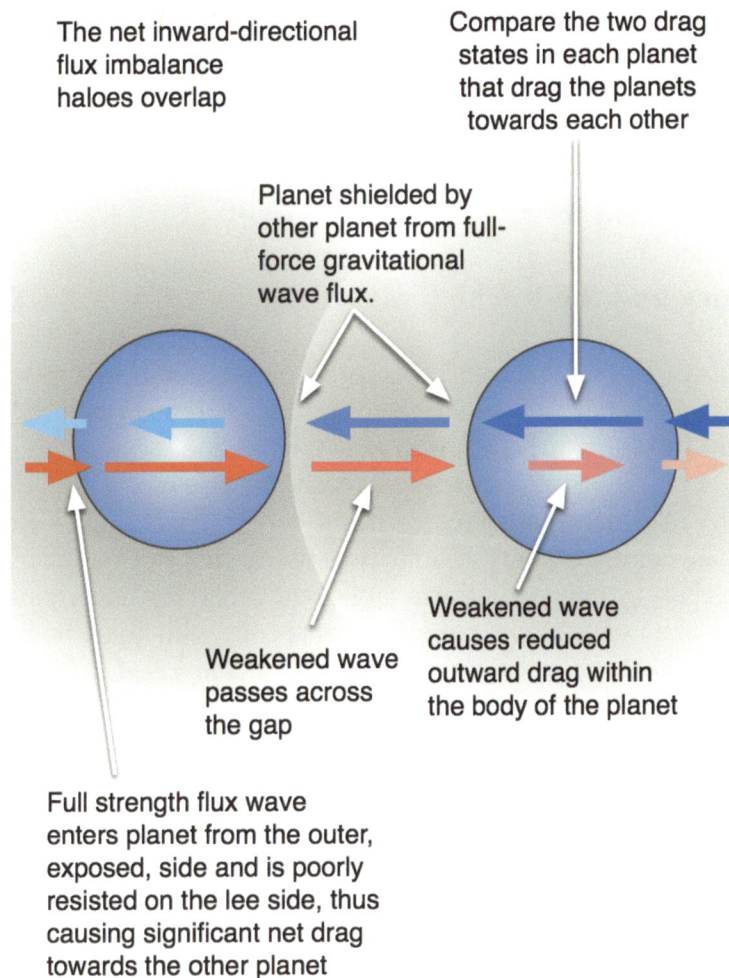

Fig 31. Artistic representation of diminution of gravitational waves as they pass through two planets.

Of course, the planets do not have to be in close proximity to each other for this mechanism to work. To realise how far a light shadow can travel before dissipating, we only need to recognise that the current search for planets around stars is conducted by measuring the fall in the star's light intensity as the planet comes between the star and our viewing point. That is a shadow travelling millions of light years. At the very same time, the orbiting planet's gravitational shadow is invisibly falling on the Earth. Thus, there is every reason to accept that gravitational shadow zones can travel immense distances, causing associated gravitational field responses.

This, according to absorption theory, is the essential mechanism by which gravity functions. There is no link between the two planets; each is driven by its own internal gravitational acceleration force. And, furthermore, it can be seen that there is no problem associated with extending this mechanism to a billion dust particles in an intra-stellar dust cloud, or to two, ten, or a thousand planets or stars. The effect is the same.

It is important to note that the DOPA mechanism is entirely and fundamentally different from the 'shadow zone' proposed by scientists many centuries ago using the then-proposed mechanism of surface pressure caused by 'aether particles' in space. They proposed that space contained an 'aether' containing moving particles that would impact on the surface of planets and cause a shadow zone between planets, thus creating a surface pressure on the outer parts of planets to drive them together. The principle was reasonable as an idea, but the engineering implications are not acceptable today because of the mechanical differences in imposed stress between this abandoned concept and this new theory of internal acceleration. These aspects of surcharge loading are covered in Section 3.10 of this book.

3.8   DOPA theory and the Roche limit

If the two planets shown in Figure 31 had been orbiting planets drawn to scale, they would be within each other's Roche Limit and would be destroying each other by gravitational instability. In that regard, Figure 31 must, clearly, be taken as diagrammatic for illustration purposes only.

DOPA theory fits in with what is accepted as being correct according to Roche's calculations. Roche calculated that there is a limiting proximity that two bodies can approach without the smaller of the two being torn apart by the gravitational pull of the other. It was, originally, an equation for rigid spherical bodies such as our Moon, but refinements cope with liquid bodies such as the Earth. *(Remember that, except for the very thin crust, our planet is comprised of liquid magma all the way down to the conjectured solid central, inner core.)*

In the following equation, the rigid Roche limit (d) is the centre-to-centre distance between a large primary planet and a smaller satellite planet at which the gravitational force on a test mass at the surface of the smaller planet is exactly equal to the tidal force of the larger planet pulling the mass away from the smaller planet. In other words, objects on the surface of the smaller planet start to float away from its surface. Of course, by this time, all sorts of mayhem are being experienced in the atmosphere (if there was one) and within the body of the smaller planet.

$$d = R_M \left(2 \frac{\rho_M}{\rho_m} \right)^{\frac{1}{3}}$$

or, equivalently, where d and R are in metres and M is in Kg

$$d = R_m \left(2 \frac{M_M}{M_m} \right)^{\frac{1}{3}}$$

Where d is the centre-to-centre Roche limit, R_M is the radius of the larger planet, M_M is the mass of the larger planet, and M_m is the mass of the smaller planet. Sigma is, of course, density—assuming a uniform density.

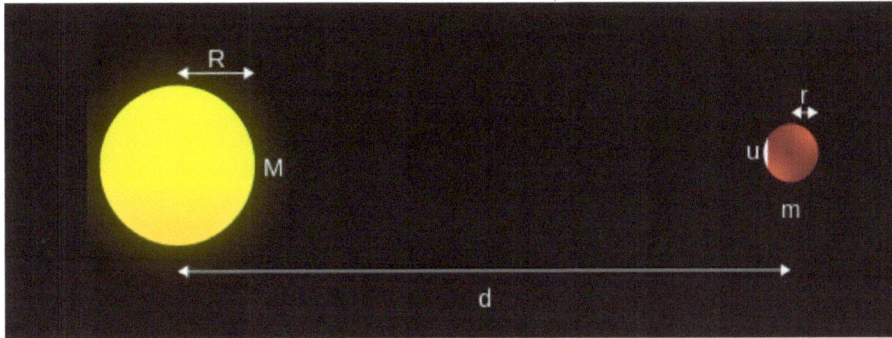

Fig 32. Dimension notation relating to the Roche limit. (Acknowledgement Wikipedia)

Figure 32 shows the dimensions involved in a Roche encounter scenario, whilst Figure 33 shows the process as a liquid satellite planet breaks up at its Roche limit. Notice that the tendency is for the small orbiting planet to become slightly oval prior to final disintegration—and if fluid, more so. These deformations are taken into account in the full Roche theory.

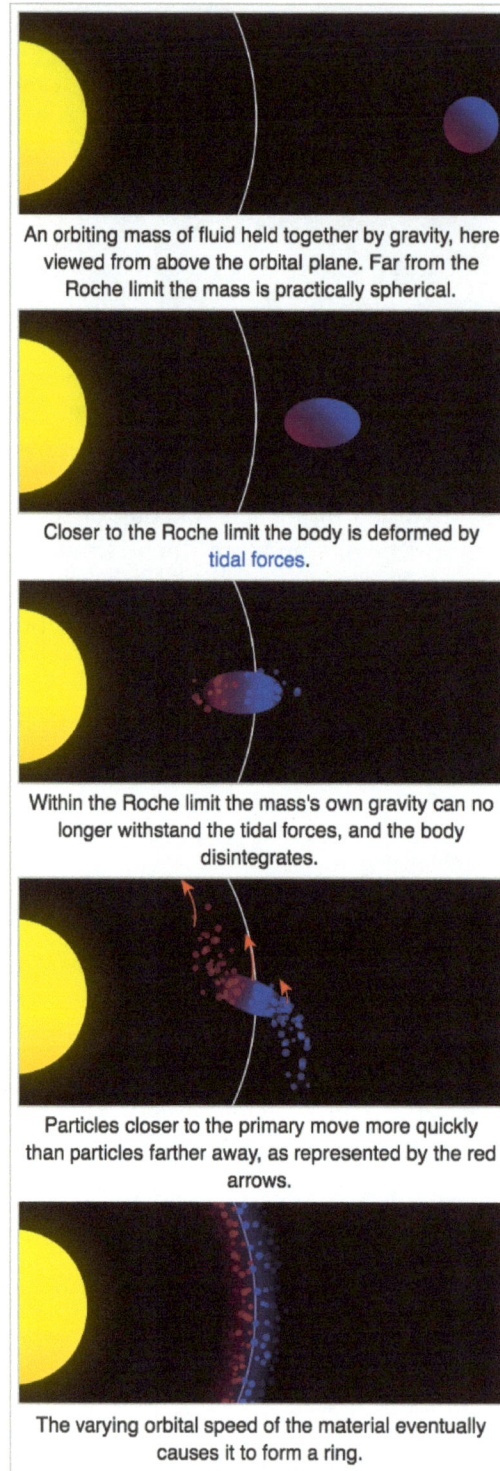

The sequence of panels shows:

An orbiting mass of fluid held together by gravity, here viewed from above the orbital plane. Far from the Roche limit the mass is practically spherical.

Closer to the Roche limit the body is deformed by tidal forces.

Within the Roche limit the mass's own gravity can no longer withstand the tidal forces, and the body disintegrates.

Particles closer to the primary move more quickly than particles farther away, as represented by the red arrows.

The varying orbital speed of the material eventually causes it to form a ring.

Fig 33. Disintegration of a smaller planet approaching a larger one and passing its Roche limit. (Acknowledgement Wikipedia)

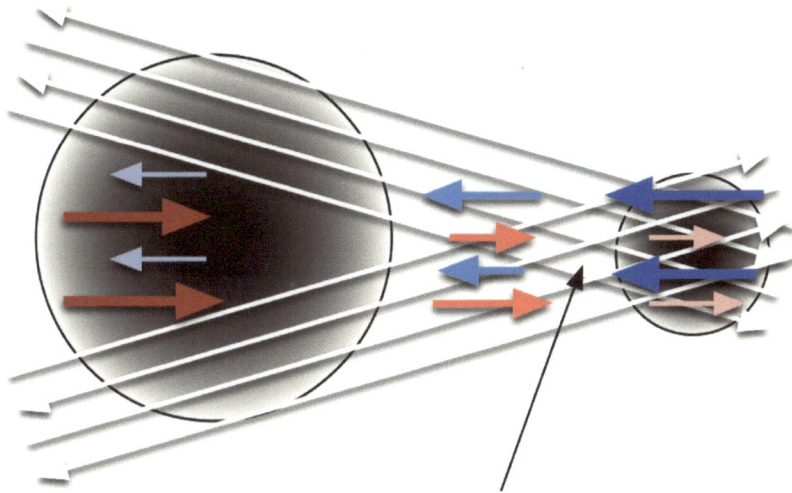

When a small planet approaches a larger one, the Roche limit is created in DOPA theory by the concentration of highly-degraded waves onto the smaller planet, whereas waves being less-degraded by the smaller planet are, conversely, dissipated across the larger one, thus reducing their gravitational impact.

Fig 34. Increasing proximity causes an excessive net gravitational force to be exerted on the inner surface of the smaller planet, towards the larger planet, leading, ultimately, to destruction at the Roche limit.

The question must be asked as to how absorption theory explains the creation of a 'Roche-type' limit based upon differential absorption of gravitational waves. Figure 34 addresses this question, showing how increasing proximity of the larger to the smaller planet occludes a greater percentage of full-strength gravitational waves from impinging in the facing surface of the smaller of the two planets thus increasing the differential absorption significantly and consequently, the net gravitational force exerted by the larger planet on the smaller. The proximity of the smaller planet does not have the same amount of impact on the larger planet. The problem with Roche's work is that it is based on the old Newtonian 'attraction' concept that has been abandoned. The Gravitational Balance Limit of DOPA theory now permits a logical replacement of Roche, based on geometric form, in the same way as its other mechanisms work.

The partial absorption diagram shown in Figure 35 primarily demonstrates qualitatively how the absorption patterns within the differently sized planets can result in an inner-Gravitational Balance Limit imposed on the inner surface of the smaller planet. It uses, as previously, hypothetical values of potential and developed gravitational force around and within the two planets. However, the following qualitative argument is sound.

Consider that, hypothetically, the incoming full-strength gravitational waves have a potential for creating a 100 m/sec^2 gravitational force in objects on the surface of either planet. This value is arbitrarily adopted, for the purposes of the diagram. Considering that traditionally the force of gravity is considered to operate from the physical centre of a body, the graphs in Figure 35, as shown, are constructed along the centre line of the geometrical axis connecting the two planets. Thus, it is a 'centre line' graph.

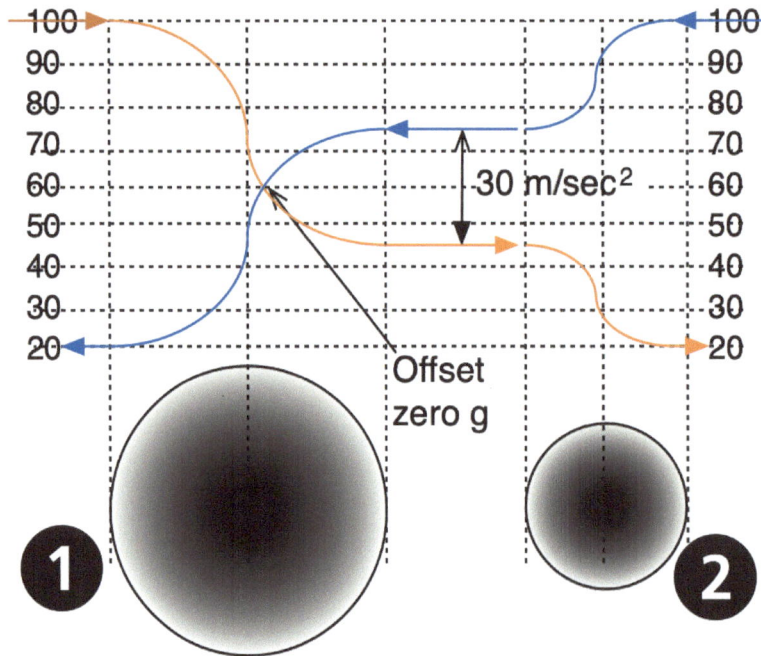

Fig 35. Differential gravitational absorption across two different-sized bodies within the Roche limit.

Referring to the graph in Figure 35, consider the left-hand side of Figure 35 (planet 1). The full-strength red-coloured wave enters the large planet with a potential of 100 m/sec^2. At the surface, (and not taking account of the counteracting blue wave) it actually exerts that much inward force immediately within the near-surface matter. But, assuming that the density of the planet increases with depth towards the centre, the gravity wave is increasingly (not linearly) absorbed by the creation of the drag within the denser matter. By the time the wave has reached the centre of planet 1's core, it has dropped in strength to having a potential of only 72.5 m/sec^2, having lost 27.5% of its potential energy. The red wave's energy level then falls a further 12.5% to the point where its strength matches that of the transiting blue wave travelling to the left—a 40% energy loss. This point is where the red and blue curves intersect and is slightly to the right of planet 1's geometric centre.

At this point, the blue wave has also lost 40% of its maximum energy and the two, therefore, balance at 60%, forming a net zero vector gravity point. In a single planet or star scenario, this would be at the physical centre of the planet, but (as may be observed in Figure 35) in a binary scenario, the zero-gravity location is physically offset as shown. This point comprises the common centre of gravity of the two bodies.

On its way out of the larger planet 1, the red-coloured gravitational wave loses a further 15% of its power and exits planet 1 towards the smaller planet 2 at a potential energy level of 45% of its original, maximum field strength value having lost 55% of its original flux strength. It crosses the intervening space gap at that level, losing no energy in the process, and enters the smaller planet only to lose a further 25% on its way through that planet. Finally, it exits '2' at a potential of only 20% having lost 80% of its full field strength.

This 80% is an immensely large, entirely fictitious, absorption figure used only for illustration purposes because it is considered that the total absorption on passing through a planet such as Earth is likely to be only a minute fraction of 1%. The potential force of full-strength gravitational waves must be immense because of the enormous mass of large stars about which we already know and which are controlled by gravity. Nonetheless, for the sake of numerical simplicity and calculation, absorption has been illustrated as being very large. Having no actual values to work with, the author has used hypothetical round figures. The reader should, please, bear this in mind.

The conditions shown by the graph concerning the net balance of force at the planets' respective outer and inner surfaces are important, of course, but are also genuinely interesting.

On the LHS outer surface of planet 1, the combined red & blue waves leave a net 80% force to the right (pristine 100 red minus degraded 20 blue). On the RHS inner surface of planet 1, the combined red & blue waves leave a net 75%L - 45%R being a 30 % hypothetical m/sec^2 acceleration force to the Left, inwards into planet 1. Thus, planet 1 experiences inwards gravitational force varying from 80% to 30% of flux maximum. The planet is stable but would be severely distorted by the variation of gravity that it experiences if such a huge differential were to be seriously proposed, which it is not. However, such vast gravitational imbalances might be possible in the case of binary neutron stars and the like.

On the LHS inner surface of planet 2, the combined red & blue waves exert their force *to the left* at the at the same 30% flux level being experienced by planet 1 on its RHS. The blue wave dragging leftwards is considerably more powerful than the red wave dragging to the right. Consequently, the surface material on the LHS inner surface of planet 2 is experiencing a destructive force whereby any objects on its surface and all of its near-surface matter will be accelerated off the planet and hurled towards planet 1 at a hypothetical 30 m/sec^2. On the RHS outer surface of planet 2, the combined red & blue waves enter the planet at a force value of 80% of flux maximum.

The differences in the gravitational force values to which both planets are subjected within the Roche zone are significant, and the reader will have noted that the geometrical effects shown in Figure 34 will affect the outcome of these qualitative observations made along the centre line of the axis of the two subject planets (or stars) shown in Figure 35.

It is most important that the reader should not assume statements concerning force at the planets' surfaces are suggesting that the gravity forces are manifesting themselves as surface pressure. Not at all. The surfaces of the planets have been mentioned because they represent distinct locations. The real accelerating effect is only manifested within the body of the planets or stars concerned.

In the case of planet 2, there is a net driving force entering the RHS outer surface inwards towards the left of 80% maximum flux. However, on the LHS face of planet 2, there is a net 30% outward-directed gravitational field that will try to drag the surface material away at an acceleration of 30 m/sec^2. This requires the necessary clarification that if planet 2 is being allowed to move towards planet 1 in differential free fall, then with the RHS face being dragged to the left at 80 m/sec^2 and the LHS face only being dragged to the left at 30 m/sec^2, the entire planet will be being accelerated towards planet 1 at 80 m/sec^2, and thus planet 2 should not disintegrate at all, but simply accelerate towards an ultimately-fatal collision with planet 1—a logical and interesting conclusion. However, if it is assumed that the two planets are actually locked in a binary orbit around each other, then the gravitational force inwards on the LHS surface of planet 2 will destroy it as the planet's surface is pulled differentially away and 'consumed' by planet 1 as planet 2 slowly spirals inwards as shown in Figure 33. This is the effect of the Roche limit.

The reader would be able to ascertain all of the above by a careful study of Figure 35, but it is hoped that the above description will make the relevant points clearer and save some time and effort.

It is of interest to note that the red waves emerge into space to the right of planet 2 at a 20% depleted level, at which point the 100% incoming blue waves 'mix' with them to create the 3-dimensional *potential* gravitational halo that was referred to earlier.

Importantly, a study of Figure 35 shows that absorption theory conforms to the requisite law that objects must 'attract' each other with an equal force. Figure 35 shows that both planets in the example attract each other with an accelerative gravitational force of 30 m/sec^2. It impinges on both of their facing surfaces. It is even more important to note that absorption theory shows *why* they do that.

The intriguing aspect that still requires some mathematical analysis is the question of when does the size of the minor body become sufficiently small that there is no Roche limit. When a small planet gets too near to a larger one, it can be ripped apart by gravitational acceleration drag forces, but a person in orbit or a feather near a planet's surface are held together by their own internal strong atomic forces and are not ripped apart. This balance between Roche forces and atomic forces is a subject not generally commented upon, but which needs examination in the context of the DOPA theory.

3.9   Absorption theory and our oceanic tides

Figure 36 reminds us of the increase in the amplitude of waves permitted to propagate when water depths increase at times of high tides. Which raises the question of what creates the tides? The answer is that the moon predominantly creates the oceanic tides that we experience on the planet Earth. The Sun also plays a minor part in helping to make tides either higher or lower than usual, depending on its position relative to the joint Earth-Moon binary axis.

Fig 36. High-amplitude oceanic breaking waves.

The concept of a common centre of gravity between the Moon and the Earth has been known for a long time and has commonly been proposed as the reason why the Earth exhibits high tides that are predominantly experienced twice a day as shown in Figure 37.

Figure 35, by virtue of establishing the reason for an offset, common centre of gravity now provides support for the concept, but, while the mechanism of rotation around a common centre of gravity is widely accepted, the popular understanding of how high and low tides manifest themselves in our oceans is fundamentally flawed.

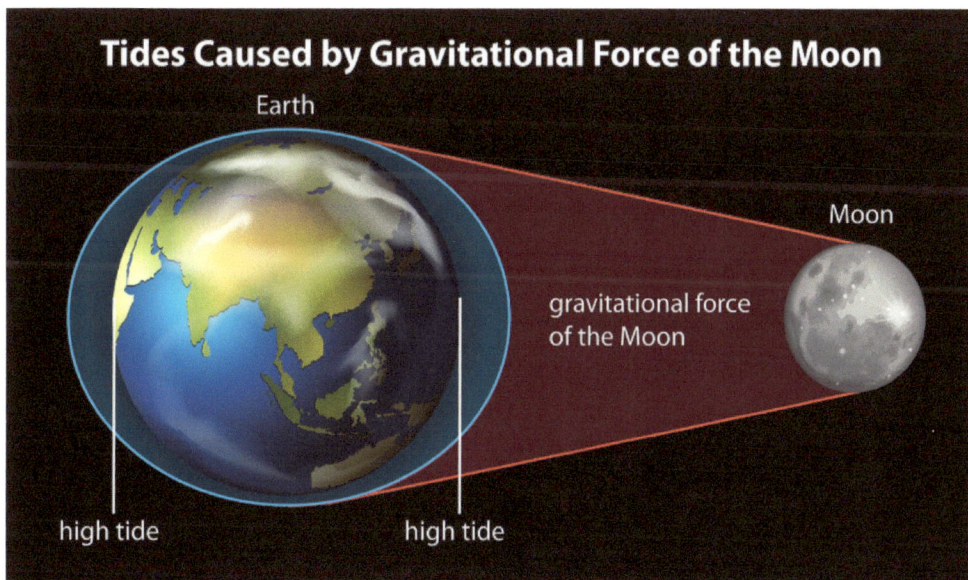

Fig 37. The Moon creates two high tides every day in most ocean basins.

If one looks at Figure 37, it seems to be an excellent diagram—nicely drawn, and showing a sort of frustum entitled, "gravitational force of the moon". Then it shows that there are two high tides on the Earth; one adjacent to the Moon and one opposite. Also, it proposes that the world's oceans and seas experience a raising of their sea level so that any particular line of longitude experiences a high tide only twice a day.

This entire concept is sadly incorrect, because it is known by oceanographers that, although the majority of tides are semidiurnal (twice a day), certain parts of the oceans experience only one high tide a day, and others, three. How can that be, if the tidal 'bulge' exists as cartooned by artists?

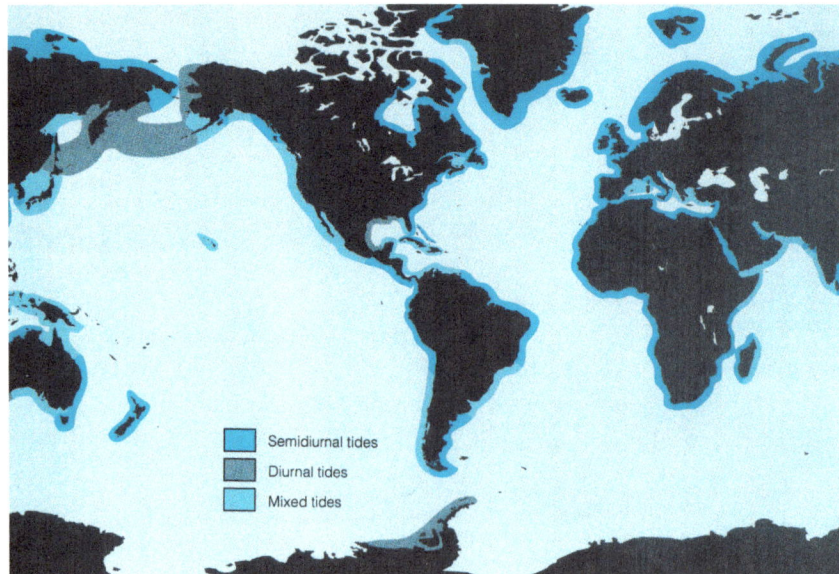

Fig 38. Different parts of the Earth's oceans and seas experience one, two, and three high tides every day. (From 'Oceanography' by T. Garrison. 1993, Wadsworth)

The answer is that the 'two-bulge' concept and drawing is false and unrealistic. The waters of the Earth's oceans do not just lift and fall as the Earth rotates, they actually have tidal wave systems that rotate around each sea and ocean basin like water in a teacup when it is stirred by a teaspoon. And, as a result of these rotational waves being formed, two things follow. Firstly, that the basins are never abandoned by high tides—high and low tides are present in each ocean basin all the time, twenty-four hours a day. Secondly, there are many points within the ocean basins where the water level never rises or falls, but remains the same permanently; these are called amphidromic points and are shown in Figure 39. There are more amphidromic points than those shown in the figure, but for the sake of clarity, a number have been excluded.

Fig 39. Tidal rotation and amphidromic points (From 'Oceanography' by T. Garrison. 1993, Wadsworth)

Figure 39 shows the lines through which high tides rotate. The points at which all of those lines meet are the amphidromic points where the water never rises or falls. This phenomenon shows that the idea of a 'bulge' is mistaken. The numbers on the figure are the hour when the high tide is present in the sea, and these record the rotation hour by hour. So, there is never a time when the Pacific Ocean, for example, is all 'high' or all 'low'. Moreover, there is never a time when the Pacific is 'high' while the Atlantic is 'low'. However, even more importantly, when arguing against the incorrect tidal concept is that the *high and low tides are present all the time in every ocean and sea basin.* They never move to match the position of the Moon or the Sun in the simplistic way of the diagrammatic models used so widely. What happens, in reality, is that the gravitational force of the Moon and the Sun 'stir' our oceans around and keep them moving. They do not lift the water in two great bulges.

When the Sun is in line with the Earth and Moon, the maximum tidal water heights in the swirls are reinforced more than usual, and we call those extra-high tides "spring tides". They have nothing to do with the spring season that follows the winter. Conversely, when the Sun is at right angles to the Earth-Moon axis, then the swirls are dampened, causing particularly-low tidal water heights around the globe. We call those low tides "neap tides".

3.10 How does DOPA differential gravity exert its force internally within bodies?

This is an important subject if one is to make valid judgments on whether or not a particular theory for the creation of gravitational force is valid or preferable.

For instance, as human beings, we associate force with external force or even pressure. If we want to push a wall over, we apply an external force to it, either by pushing it on its side with our hands or by applying an external force using a machine. If we want to drive a pile into the ground, we apply a repeating load to the top of the pile using a piling hammer. If we want to hold up the dome of a cathedral, we construct pillars, and the dome forms a surcharge load on the pillars. These are all examples of our collective perception of load and force.

There are other aspects that professionals know about but people do not think about in detail because they are lay-persons in terms of civil engineering and have not been trained in that field. However, we are mostly aware of the fact that if we go down under the surface of the sea, the pressure of the water becomes greater with depth, as shown in Figure 40.

Water weighs 1 tonne per cubic metre and, therefore, for every metre that one goes down under the surface of the sea, the pressure increases by an additional 1 tonne per sq.m.

At just three metres down, the pressure on one's ears, nose, lungs, and body is 3 tonnes per square metre. It does not sound like it is very deep, but a person needs training to dive even to that shallow depth in safety, primarily because of the gravity-induced pressure.

We are all aware that bathyscaphes are, like submarines but more-so, designed to resist the high pressure of water at depth, to stop the water pressure crushing them. Alternatively, in the opposite case, we are aware that if we take a lift to the top of a tall tower or building, our ears will 'pop' because the air pressure is less higher-up than lower-down. Also, we have heard that if we go high up enough, the air becomes so thin that it stops existing altogether at a great height and becomes space.

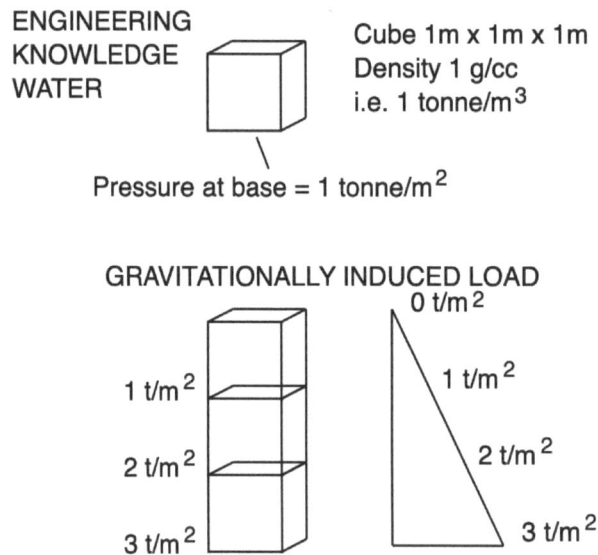

ENGINEERING KNOWLEDGE WATER

Cube 1m x 1m x 1m
Density 1 g/cc
i.e. 1 tonne/m^3

Pressure at base = 1 tonne/m^2

GRAVITATIONALLY INDUCED LOAD

1 t/m^2

2 t/m^2

3 t/m^2

0 t/m^2

1 t/m^2

2 t/m^2

3 t/m^2

Fig 40. In water, pressure increases linearly with depth.

As lay persons, we think, "it is the weight of the atmosphere above that causes the air to be denser and under more pressure lower down," or, "it is obviously the weight of all that deep water that creates the huge pressure at the bottom of the ocean". We would be quite right in thinking and saying that. So, what is the problem?

There *is* no problem as such, but what people do not think about is how gravity exerts a force on water and gas to create that pressure? A workman cannot just push it down with his hand like he can with a wall. In fact, one cannot pressurise gas or water by a simple external force unless they are in a sealed container—and the atmosphere and the oceans are not sealed containers. If we sail a huge container ship on the ocean, it does not increase the pressure at the bottom of the ocean at all; the water just moves around it. So, how does gravity 'pressurise' water and air?

The answer is that gravity tries to accelerate *every single atom* individually downward towards the centre of the Earth. There is no other explanation needed. And, that tells us something—in fact, a lot—about how gravity operates on solid objects.

Gravity does not just squash water and gas down from the top as if one placed a hand on a person's head and tried to push them down. Gravity pulls them down by acting on every atom in their physical volume.

But, how does it do that? Until now, no one has put forward an acceptable theory. However, DOPA theory addresses this and explains it using just the four basic assumptions (tenets) set out and justified at the very start of this book.

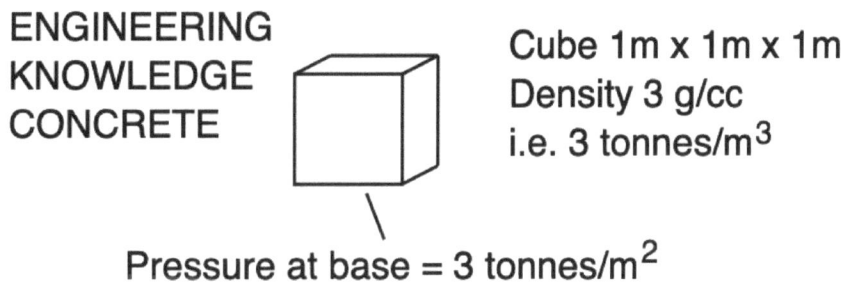

Fig 41. An ultra- dense 1 cu m concrete cube weighs 3 tonnes and therefore generates a pressure at its base of 3 t/m².

Let us, in Figure 41, consider an ultra-dense concrete cube measuring 1 x 1 x 1 metres, and let us say that it is made from a super-dense concrete weighing 3 tonnes. If asked why it weighs 3 tonnes, most people would say, "It weighs 3 tonnes because it contains 3 tonnes of concrete!" which is an unfortunately-circular argument. What they do not think about is exactly how gravity exerts its force on the concrete cube to create the 'weight'. After all, what is weight other than the downward force exerted by a mass of material? In the case of the cube, it weighs 3 tonnes on any crude scales, and because it has an area at the base of 1 sq. m, it exerts a pressure on the ground of 3 tonnes per sq. m., which we write as 3 t/m².

So, how does gravity exert its force on solid matter such as concrete? In the same way as we realised with air or gas, gravity exerts its force by accelerating every atom of the concrete downward towards the centre of the Earth. So, it does not do it by just pressing down on the top with a surcharge load (like patting a person on the head).

Can we make a guess as to what is happening inside the block of concrete as the result of the atoms being individually accelerated down by gravity?

First of all, let us look at whether the concrete deforms. Earlier, we discussed that a planet in free gravitational acceleration would not deform even though vast and massive. However, a block of concrete standing on the ground will deform elastically because it is not in free-fall acceleration. It is being subjected to a potential-acceleration stress but is being restrained by the ground beneath it. The downward stress on the chemical bonds and atomic bonds in the concrete causes them to squeeze down a little like a spring. At the same time, they squeeze outward like a sponge does when squeezed down from above. So, the distortion experienced by the concrete cube, although minute, is measurable and comprises a vertical shortening and a lateral widening.

It should be noted that, if we were to compress the concrete cube from above, we would get a similar, but not identical, deformation. This is explained below.

Alternatively, if we drop the concrete from the top of a tall crane, once released and accelerating free, it will not become distorted and will recover from any distortion induced by being supported on the ground or by being hung from the crane hook. That is accepted engineering knowledge, and DOPA theory conforms to it.

So, we can see that there are distorting stresses in the cube of concrete because it is being forced downward by gravity and kept stationary by the support of the ground. To look at what this means in terms of stress distribution, we could try to put our hand under the cube by lifting one corner, but we cannot do it because the cube weighs 3 tonnes—more than a large car. So, if we were to use a crane to lift the cube and put it on our hand, we would experience the full pressure of 3 tonnes/m^2. Our hand would be crushed.

However, let's say (as in Figure 42) we cut a thin slice across the top with a laser cutter. Then, if we tried to lift that thin slice, we would find that we could do it. "Of course!" you would say, "The slice naturally weighs less because it is smaller". So you would not be surprised if, when you put your hand under the thin slab it would not be crushed. A bit heavy, but OK.

And when you consider that, you would realise that the block is made up of a considerable number of those thin slices.

**What if we slice off a
thin layer at the top of
the cube, rather like a
very thin paving slab?**

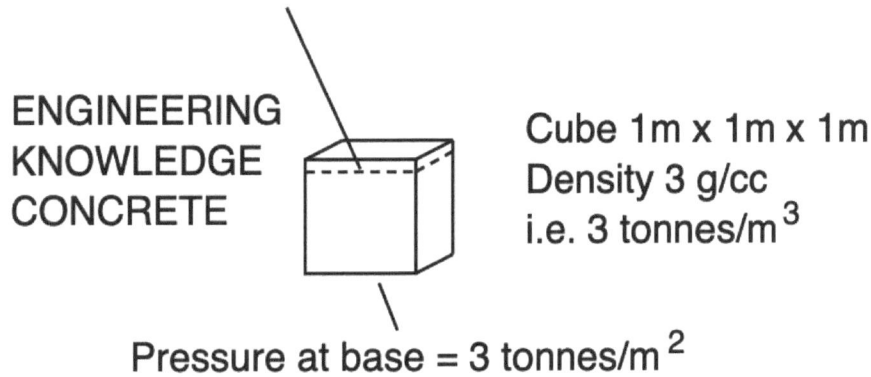

ENGINEERING
KNOWLEDGE
CONCRETE

Cube 1m x 1m x 1m
Density 3 g/cc
i.e. 3 tonnes/m^3

Pressure at base = 3 tonnes/m^2

Fig 42. A small slice of concrete does not weigh as much as the cube itself.

What is that teaching us? That the pressure in a solid concrete cube varies—just like in a liquid. If we insert our hand near the top, the block exerts a smaller downward pressure than if we put our hand in at the bottom. And that would lead us, as engineers, to expect that the lateral distortion of the cube will be greater nearer the base than near the top when loaded by gravity. And if we measured it, we would find that this was correct.

Let's now look at a more interesting case. Let's take a column of concrete measuring 1 x 1 metres on the sides, by 3 metres tall. Its basal area is the same as the cube, but it is three times higher, as shown in Figure 43.

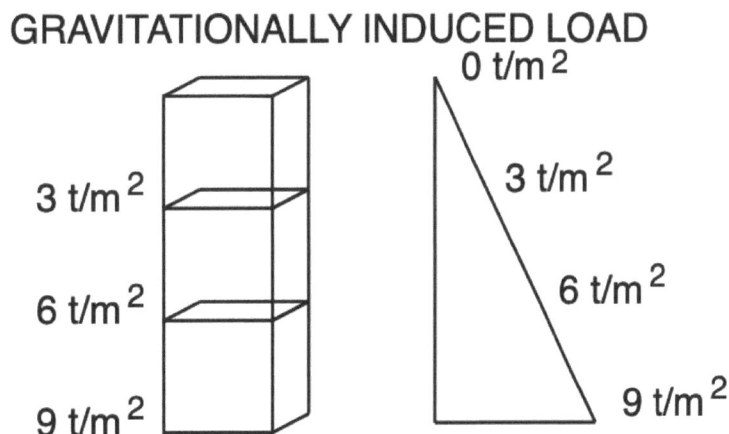

GRAVITATIONALLY INDUCED LOAD

0 t/m^2

3 t/m^2

3 t/m^2

6 t/m^2

6 t/m^2

9 t/m^2

9 t/m^2

Fig 43. A 3-metre-high ultra-dense column of concrete weighing 9 tonnes.

Naturally, since 1 cu.m. weighs 3 tonnes, 3 cu.m. will weigh 9 tonnes, and the pressure on the ground at its base will be 9 tonnes per sq. m. However, what about within the column? What happens to the pressure? Is it numerically identifiable? The answer is 'yes'.

If we cut the column into thirds and put a pressure metre at two levels as shown in Figure 43, they will read 3 t/m^2 and 6 t/m^2 respectively. So, we can conclude that, just as with water, gravity is grabbing each individual atom in the column and trying to accelerate it downward such that the internal pressure is increasing downward within the column. This internal way of loading is an engineering aspect that people do not recognise and even civil engineers who do recognise it do not ponder upon why or how that is achieved. Mostly, people just think that, somehow, gravity just kind of 'grabs' the column and pulls it downward. But our slicing experiments in Figures 42 and 43 have taught us that this is not the case.

What we can observe by slicing the column into thinner and thinner slices is that the gravity-induced pressure within the column increases linearly with depth—as shown in the graph on the right-hand side of Figure 43.

Why does the pressure increase downward if every atom is being dragged down individually? The answer is because every atom in a standing object is not being accelerated individually; each atom is trying to be dragged down but is resting on every other atom, which is resisting it and adding its own weight to the atom below, which is itself being dragged down. And so, there is a cumulative effect all the way to the base. The elastically-distorted shape of the column is being maintained by the forces that exist between every atom in the concrete. The atoms form a solid mass that is held together by strong chemical and nuclear electromagnetic forces. Those forces have nothing to do with gravity.

From all of the above, we can recognise that gravity tries to move every atom whether in a gas, liquid, or solid.

Today the very logical, but incorrect, conclusion that people still make, which is natural, is that if everything is being 'pulled' towards the centre of the Earth by gravity, then, somehow, the Earth is making that force which is pulling everything down. And, that is what Newton is reputed to have conceived when he saw the apple drop from the tree to the ground. He thought that the Earth had pulled it down, which is what led to his concept of gravitational attraction. Einstein disagreed with that, and so does Roberts. Gravity is not an attraction. It is an internal acceleration created within each object independently. According to absorption theory, the Earth does not attract the apple, and the apple does not attract the Earth. They are both driven together internally and independently by the process explained and published in this book for the first time.

So, the ultimate questions are, "what is gravity?" and "how does it grab every atom?" Absorption theory provides the answers to those two critically important questions.

Moreover, within the Earth itself, gravity performs the same procedure as in the concrete column, with pressure increasing steadily downward within the crust and body of the Earth right down to the centre of its core.

To finish this discussion, we need to understand a little more. We need to also learn about what happens if we put a load on top of a column (which we call a surcharge). Engineers know about this as well. What will happen to the internal pressure then? Will it increase from top to bottom in the same way as a gravity-induced load? Figure 44 shows us a column of lightweight polystyrene with a 1 tonne surcharge placed on top of it. It shows that a surcharge-induced load is entirely different from a gravity-induced load. The induced pressure in the column does not increase from top to bottom but stays the same.

Practically speaking, the vertical and lateral deformations of the column are evenly distributed from top to bottom. (We are, for the purposes of this exercise, ignoring the very lightweight and minute self-stress of the polystyrene column itself.)

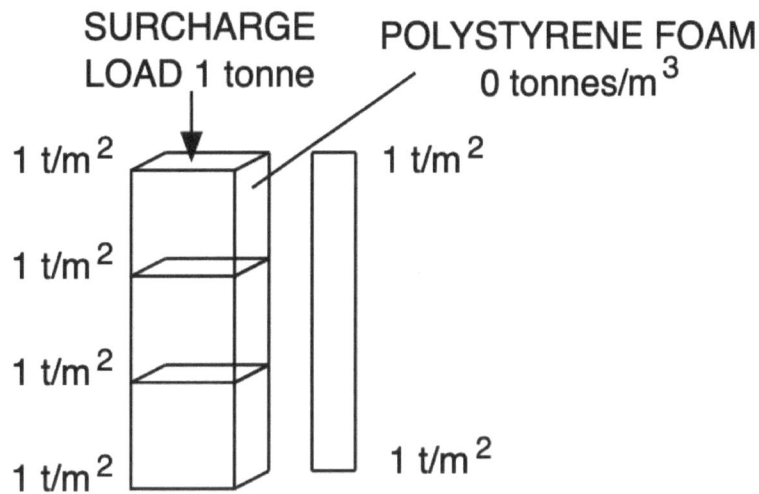

Fig 44. A 3-metre-high column of polystyrene with a 1 tonne surcharge.

One can see from the graph at the right-hand side that the weight of the surcharge and its induced pressure of 1 tonne/m^2 is carried down uniformly within the column. Assuming that the polystyrene effectively weighs nothing, then our pressure metres placed at intervals will all read the same, and the pressure on the ground will be 1 tonne/m^2.

There are some interesting and relevant consequences to the difference of surcharge and internal pressure loading. To see these, look at Figure 45.

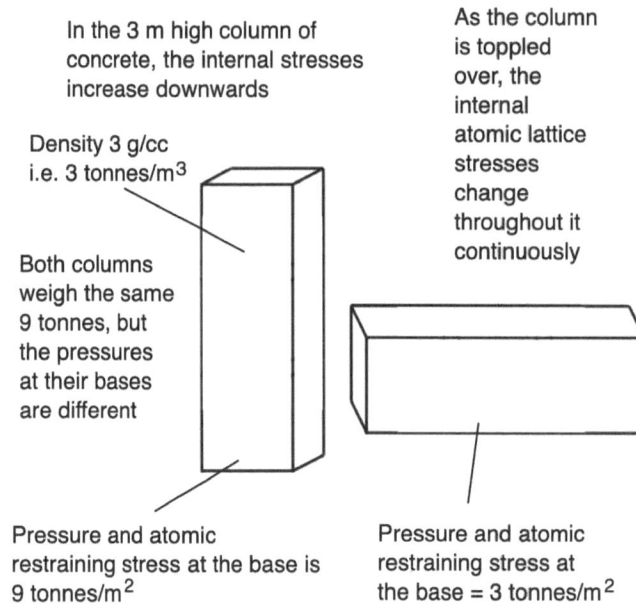

In the 3 m high column of concrete, the internal stresses increase downwards

As the column is toppled over, the internal atomic lattice stresses change throughout it continuously

Density 3 g/cc i.e. 3 tonnes/m^3

Both columns weigh the same 9 tonnes, but the pressures at their bases are different

Pressure and atomic restraining stress at the base is 9 tonnes/m^2

Pressure and atomic restraining stress at the base = 3 tonnes/m^2

Fig 45. Different pressures and stresses in a column of concrete when rotated sideways.

When the column is vertical, it weighs 9 tonnes and exerts a pressure on the ground of 9 tonnes/m^2. Yet, if, as shown in Figure 45, we rotate it sideways and place it on the ground, it still weighs 9 tonnes, but it only exerts a pressure on the ground of 3 tonnes/m2. This means that the pressure within the concrete now is the same as a single cube of concrete weighing 3 tonnes. It has, for all practical purpose, been converted into three cubes of 1 tonne each. What has happened internally?

The atomic lattice held together by strong atomic forces had to restrain a pressure of 9 tonnes/m^2 when the column was vertical, but, as it is rotated, gravity only grabs a third of the atoms vertically and so relaxes its pressure at the base and the stress on the internal atomic lattice whose distortion reduces accordingly. The whole process is logical and automatic.

Once it has been shown what gravity is and how gravitational waves drag the individual atoms, the whole picture is complete, **and we can, at last, understand gravity.**

We can conclude, then, that force downward by surcharge pressure is fundamentally different from force downward created internally by gravity.

Which helps us to understand the effect on cosmic bodies such as the Earth when absorption theory proposes that it is gravitational waves that interact with each atom creating a minute drag on each, accelerating them in the direction of *dominant* gravitational wave propagation. This book has explained why, in the case of a planet, the dominant direction of wave propagation is downward.

This book demonstrates that the case of a cosmic body such as the Earth or Moon is simply a scaled-up version of what we can observe and measure in a concrete column.

And gravity has an effect on us as human beings. Figure 46 shows a diagram of a human male who supports his weight through his skeletal frame.

54

ENGINEERING
KNOWLEDGE
HUMAN

GRAVITATIONALLY INDUCED LOAD

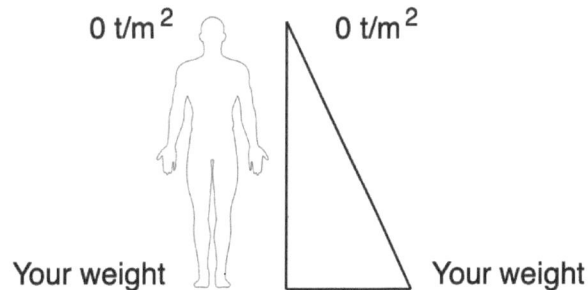

$0\ t/m^2$ $0\ t/m^2$

Your weight   Your weight

THE LOAD WITHIN YOUR BODY FRAME
INCREASES STEADILY DOWNWARDS AS
GRAVITY DRAGS YOUR ATOMS TOWARDS
THE CENTRE OF THE EARTH.

BECAUSE YOUR ANKLES ARE VERY LOW
DOWN THEY ARE HIGHLY STRESSED, BUT
YOUR NECK IS ONLY STRESSED BY THE
WEIGHT OF YOUR HEAD

Fig 46. Human skeletal stress increases downward.

We do not think about it, but there is zero vertical gravitational load at the very top of our heads. Our necks support only the moderate weight of our heads, and the pressure on our spines increases downward, which is an additional reason why (apart from walking on our hind legs) we suffer from lower back pain. The stress increases in our knees, and our ankle bones take the highest pressure of all. They carry our entire body weight through them. So, we can use our discussions about the concrete column to recognise that the same thing is happening to us.

We are not being *pulled* down by the Earth, but the small difference between the incoming gravitational waves that are passing down through our bodies all the time, and the upcoming, depleted, gravitational waves that are travelling upward through our bodies, leads to a slight difference in which a net force is trying to accelerate our atoms downward at the rate of about 9.81 m/sec^2. The downward waves accelerate our atoms downward just that bit harder than the upward waves try to push us up into the sky, and so we stay firmly planted on the ground—except when we run or jump, in which case we are soon forced back down again.

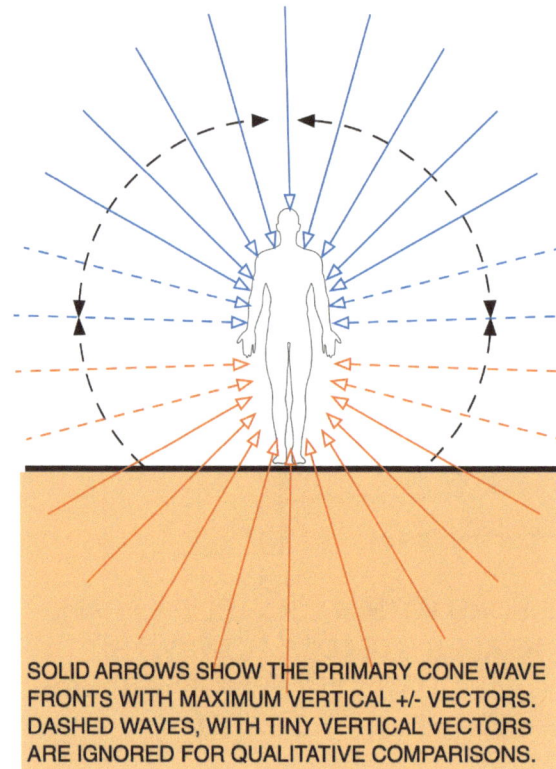

SOLID ARROWS SHOW THE PRIMARY CONE WAVE
FRONTS WITH MAXIMUM VERTICAL +/- VECTORS.
DASHED WAVES, WITH TINY VERTICAL VECTORS
ARE IGNORED FOR QUALITATIVE COMPARISONS.

Fig 47. Gravitational waves entering the body from above (pristine, blue) and below (degraded, red).

Note, in figure 47, that although the waves are shown just outside the body of the man, they are not pressing on him from outside. They pass into him and through him, and while passing through him, they generate the net gravitational force and acceleration that we call gravity. Their passage through the person and onward has been omitted for clarity.

The balance between all of the downward and upward gravitational waves results in a single vertically-downward net vector which presses our body down towards the centre of the Earth. We feel our weight, and we do not think about it when we weigh ourselves in the morning. We would think about it very deeply, however, if we were to step off the edge of a tall sky-scraper building.

The battle between the incoming and outgoing waves is eternal. Luckily, it is totally reliable; otherwise, we could be crushed to death instantly—which, of course, is a danger we shall all have to face once we start to control gravity for human engineering and domestic purposes.

So, it is now possible to recognise and accept what has been demonstrated and what all engineers know is true: that gravity is an acceleration exerted within physical structures, feathers, humans, and cosmic bodies alike.

CHAPTER 4 - A GOOD THEORY SHOULD ANSWER ALL RELEVANT QUESTIONS

Whereas **these questions do not form a part of the theory,** they are provided here to demonstrate that the author has given thought to explaining where his working partial absorption mechanism fits in with what we observe. The answers provided in this section are developed by considering the proposed mechanisms of DOPA theory and asking what would be the logical consequences in the context of each question. This is a stage of reasoning higher than speculation, because these answers logically and inexorably follow the acceptance of the tenets and mechanisms of the theory.

4.1   How does the theory explain why interstellar dust clouds form into planets and stars?

Omnidirectional gravitational waves passing through a dust cloud generate multiple gravitational shadows between particles that create a central zone of low gravitational potential towards its centre. Dust is, therefore, accelerated inwards towards that central shadow zone. The more the dust coalesces, the stronger the inward drag becomes as the density of the mass increases, creating an increasingly rapid accumulation and aggregation of material. This is a geologically-slow process, taking of the order of billions of years to create a solar system with planets.

4.2   How does the theory explain how gravitational waves create a force that tries to make large objects such as planets and stars spherical?

It explains that omnidirectional waves are broadly equally partially absorbed from whichever direction they come through any given planet, and so the in/out differential is broadly the same everywhere within the body. Thus, a spherical shape is formed.

Yet we have to recognise that, on the smaller scale, small differences of absorption caused, for example, by the presence of very dense ore bodies in the Earth's crust can lead to detectable differences in the gravitational field. These can be measured by ground instruments or very sensitive instruments carried in survey aeroplanes.

4.3   How does the theory explain why the same force-creation mechanism that creates a star does not crush a small object such as a human into the shape of a sphere?

Consider an object such as a human body floating in a space suit far away from any planet, thus being in a state of effectively-zero gravity. In such a case, the answer is just that the relative differential gravitational forces created within such a body (that would try to turn it into a sphere) are so small that the atomic and molecular bonds which hold such solid matter together are too strong for them to break or even noticeably distort.

A feather on Earth, for example, has such a low mass and density that it absorbs almost none of the differential gravitational waves passing through it, both up-going and down-going. Thus, the force that tries to drag it down to the Earth is very small, directly in proportion to its mass, and this force is insufficient to break it apart into its constituent atoms. So, when we drop it, all we see is that the feather floats gently through the obstructing air, to land gently on the ground.

Until an object gains the size and mass of a moon or planet, there is insufficient absorption of gravitational waves to generate a force great enough to squash it into a spherical shape. We can see photographs of large comets that are passing by our Earth. They have even been visited by spacecraft and photographed. One can see that they are large, but still not large enough to be forced into a spherical shape. They are always quite irregular, which conforms to partial absorption theory.

4.4   How does the theory explain why potential gravity zones are created around cosmic bodies that reduce in intensity away from those bodies?

The theory proposes, in its tenets, that 100% full-strength flux waves are ubiquitous. These co-exist with the depleted exit waves that have passed through any existing bodies. The amplitude differences in the co-existing waves at all points form a three-dimensional zone of potential gravity that is utilised by any matter located within it. This is the universal gravity field, and this is how it is created. This concept is unique to DOPA theory.

Inevitably, the presence of the spherically-outward-travelling 'exit waves' surrounding any planet creates the potential gravity zones that are commonly called 'gravity wells'. Matter does not 'slide' down the gravity zones (as postulated by Einstein), it is moved by internally-created forces applied to every atom; the pre-existing differential wave flux field creates the force within matter.

The proposition that each atom is individually dragged means that, under unhindered conditions, the applied force develops a true gravitational acceleration whereby, like a person in freefall, the planet does not 'feel' that it is being 'pushed', and therefore, there is no induced distortion as it moves itself towards another planet or star.

If a planet is allowed to follow its accelerated path, then it will be in true gravitational free fall and will remain spherical, but if it is being held away from another planet by virtue of orbiting it and thus experiencing a counter-balancing centrifugal force, it will not be able to respond to the imposed forces on its atoms and it will thus distort minutely, usually into an oblate form, especially if, like the Earth, it is a primarily liquid planet. This is why Figure 33 shows an oblate satellite planet being destroyed gravitationally at the Roche limit.

4.5   Does the theory explain how gravitational force applies itself so that satellites can maintain a circular or elliptical orbit around a planet?

Before addressing this question directly, we need to look at some of the peripheral aspects of bodies in orbit around planets.

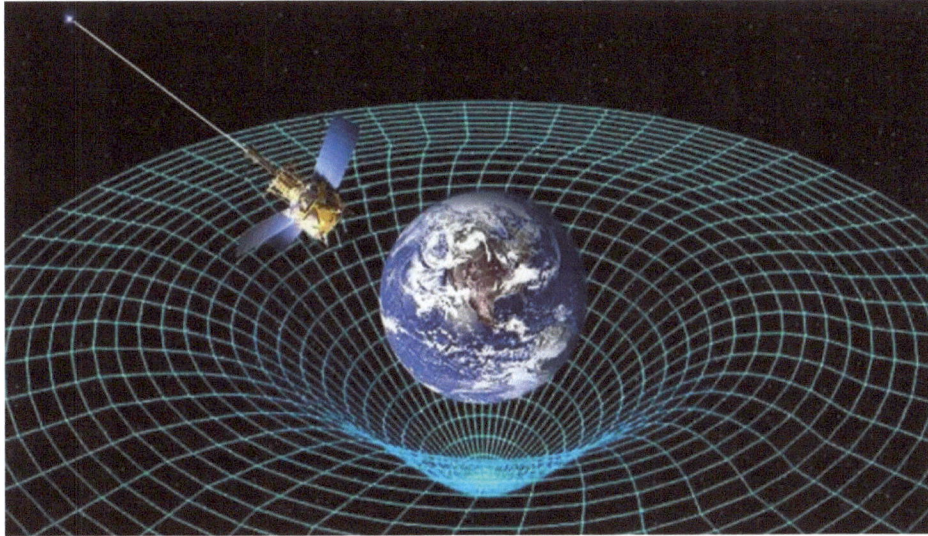

Fig 48. Gravity well with incorrect mesh shape displayed.

Figure 48 shows one popular "artist's impression" of a 'gravity well' which is drawn in this way because of the misnomer of its name.

It is an artist's graphical representation of how Einstein's 4-dimensional spacetime is distorted by the presence of the matter contained in a planet such as Earth. In this interpretation, the increase in gravitational force created by the planet is shown as a vertical depression downward, thus creating the analogue of a 'well'. The more depressed the mesh is, the more distorted spacetime is supposed to be, creating a greater 'incentive' for any nearby matter to slide down the distortion towards the planet. This distorted spacetime is Newton's equivalent to a gravitational force field.

Unfortunately, the diagram has some errors and is, consequently, misleading. These errors are created because it illustrates a distorted 2-dimensional mesh. One can forgive the difficulty of illustrating a four-dimensional phenomenon on a two-dimensional piece of paper, but not technical errors.

There are two things that are not so forgivable in the diagram.

Firstly, there is the rising up of the mesh as it approaches the planet before it drops into its 'well'. That never happens, because that would mean that the planet's spacetime/ gravitational force was at a minimum outside the planet, somewhere in space, which would prevent any other body from moving towards the planet if it were outside the minimum zone. The strength of the spacetime gravitational field reduces steadily away from any planet until it levels out at zero at some considerable distance, or until it approaches another cosmic body, whereupon it starts to fall again to describe the increase in gravity associated with that body.

The second thing is that the net gravitational force does not continue to become a maximum at the centre of the planet, as shown. The maximum gravitational field for a planet or star is at its surface. Within the body, the gravitational vector forces decrease to zero at its centre. If the diagram were even moderately accurate, it would show the mesh increasing to zero at the centre of the planet,

59

which would mean that the mesh should rise up to the external base level, not forming a 'well'. The intersection nadir of the mesh would be exactly at the surface of the planet. In the case of spacetime, it is not clear how the artist's well represents Einstein's concepts, but it surely cannot be a maximum at the centre of the planet, as indicated in the diagram. It seems a shame that whoever commissioned the artwork did not take the trouble, or meet the cost, of ensuring that it was as reasonably technically correct as it could be. The lesson to be learnt from this is that the reader should take any such mesh diagram with a large pinch of salt.

If we consider the following diagram, there are two main differences between figures 48 and 49.

The first is that figure 49 is not intended to represent Einstein's spacetime, it is intended to represent Roberts' potential gravitational field. In other words, the intensity of shading represents the maximum potential strength of any gravitational force generated within any matter that might enter that field.

And, secondly, the following Roberts diagram is better, in that it represents a 3-dimensional globe that weakens with distance from the planet and also decreases internally within the planet because it represents gravitational force and not the bending of spacetime. This diagram conveys a much more straightforward concept.

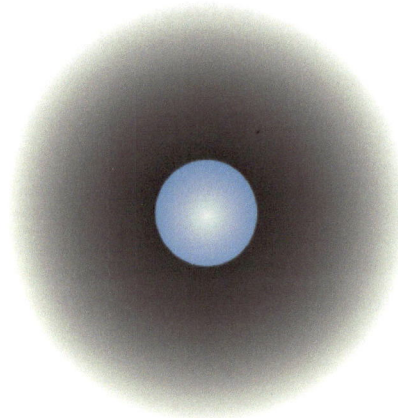

Fig 49. The potential gravitational field diminishes outward asymptotically more rapidly than Newton's inverse square law. The field also decreases inwards within the planet according to Roberts' shell theory.

It is proposed that Figure 49 represents, more realistically, the three-dimensional potential gravitational force field around a star or planet and even within it—which is something that Figure 48 does not do at all. The outer, grey, sphere represents the potential gravitational field diminishing with distance away from the spherical body; the inner shading (blue) represents the potential field diminishing towards the centre of the spherical body to a net zero value at the centre. Neither Figure 48, nor Einstein's spacetime represents the gravitational potential within a planet or star satisfactorily in this way.

It is an essential part of DOPA theory that in the vacuum of space, there is no actual 'force field'. It is a *potential* force field that exists as a potential by virtue of the co-existence of pristine un-depleted gravitational waves with depleted ones. When an un-depleted wave is counteracted by an oppositely-moving depleted wave *within any matter*, then the difference between the two results in a net gravitational, acceleration-inducing, drag within that matter. Thus, either within space or within matter, the difference between opposing wave strengths (amplitudes) creates a potential for gravitational force to be developed in the form of atomic drag. It is only *within* the matter that the net difference between the waves is realised and turned into a net force in the dominant wave-travel direction.

Gravitational force can only manifest itself inside matter that is present within the potential field, and even then, it is not a 'link-up' with other matter, it is matter that is subject to a logical, independent, automatic attempted acceleration towards any nearby body of matter. Physicists view the internal 'force' as an 'acceleration', but for the purposes of this book, the author has tried to differentiate where possible between the two such that the term 'drag' implies the force and 'acceleration' refers to the resultant increasing velocity along a given vector if the force is not resisted.

Insofar as that is concerned, the outcome of this part of the theory (the creation of *gravity zones*) is virtually identical to that of Einstein's curved spacetime. After all, the Moon did not change its orbit when spacetime was accepted as a replacement for Newton's universal law. The Moon will not change its orbit if the author's theory is correct, either. *The difference is in the representation only*, in that this theory relates to a simple 3-dimensional field created by a mechanism that is simple, logical, described, and explained; which contrasts with the lack of explanation for the existence of spacetime and the lack of reasoning by Einstein as to why matter should slide down his proposed spacetime wells.

In the case of a star, every atom or speck of dust ejected from a star's surface has to pass through the potential gravitational field (the star's own potential gravity sphere) and, while within it, in addition to the star's magnetic pull, the differential gravitational wave field exerts its net force inwards, back towards the star's centre. That is what we have always called "gravity". Absorption theory works neatly and fits into all scenarios.

It is not difficult to see that, if every atom and speck of matter is affected or controlled by the potential force field, then it will be either dragged directly back into the star, as shown in Figure 50, or may become part of a zone of orbiting matter that resists being accelerated inwards by virtue of its high orbiting velocity. Mostly, those particles will be part of a vast solar atmosphere that ultimately extends out for hundreds of millions of kilometres from the Sun in the form of the solar wind.

Fig 50. Solar flares being dragged back into the Sun by magnetic loops and gravity.

In absorption theory, when an object is in orbit around a planet, as shown in Figure 51, it has no direct, fixed tie to that planet. Instead, as it moves in its orbit, it does so at a balance radius within the outwardly-reducing net potential force sphere.

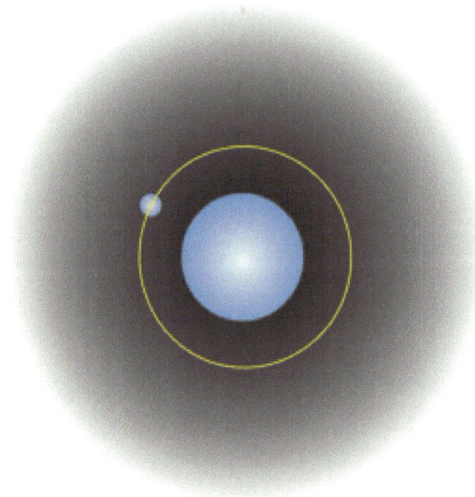

Fig 51. A satellite in circular orbit around a planet.

Homing in on the subject of the question, it is the author's proposal that the Newtonian law controlling the force of attraction between matter contains an unrecognised flaw relating to the mechanism that keeps a satellite in orbit around a planet or star. The theory is flawed in that it works for the wrong reason.

Newtonian theory has (for lack of anything better) always been used to explain how the centripetal force of gravity holds a satellite at a fixed distance. In particular, Newton's law states that the gravitational pull of the planet depends upon the inverse square of the distance between the centre of the planet and the satellite's centre of gravity. However, that concept contains a grave flaw of which no one appears to take any notice, or else, which no one has realised.

The flaw is that a Newtonian gravitational force holding a satellite in orbit at a particular distance is a balance point where its centrifugal force balances the gravitational pull at that particular distance in which the force varies inversely with the square of the distance. Even Einstein's gravitational wells conform to this principle, having spacetime distortions that conveniently emulate that behaviour.

It is known, from standard physics, that a body in a rotational orbit continually tries to move away from its controlling planet along a tangential path. It is prevented from doing so by the centripetal force of gravitation which is itself (according to Newton) dependent upon the inverse square of the distance between the centre of the planet and the centre of gravity of the satellite as shown in Figure 52.

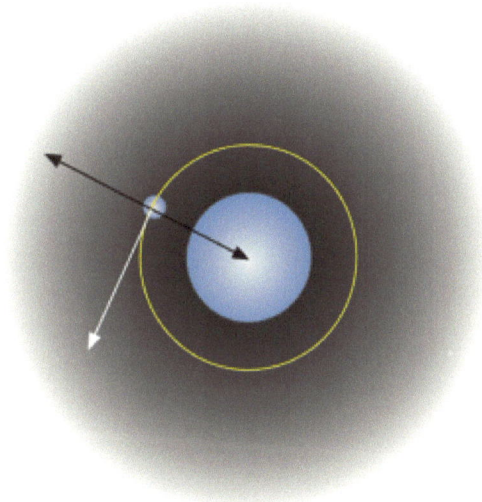

Fig 52. A satellite attempts to follow a tangential escape path away from the planet.

However, when, for example, a relatively small satellite tries to continue travelling on a straight course and therefore tries to move tangentially, it necessarily, tries to move slightly further away from the planet, as shown in Figure 52. Yet we know that a satellite, by virtue of its initially-imposed orbital velocity can hold an orbit at a very particular distance away from the centre of the planet in order to conform to the law of conservation of angular momentum, or equivalently, Kepler's second law.

The weakness in Newton's theory in this context, therefore, is that, as the satellite moves slightly outward in its orbit, in order to travel tangentially, its distance from the centre of the planet increases and therefore, the force holding it to the planet must decrease. This should lead to a cascading reduction in the gravitational centripetal force that holds it to the planet. And yet this does not happen, in practice.

Let us look again at this, but this time iteratively. Firstly, the satellite is at a distance d from the centre of the planet. It tries to move a microscopic distance Δd on a tangential path away from the planet. So, again microscopically, the radius away from the planet becomes $d+\Delta d$ metres. Since, according to Newton's equation, F is inversely proportional to d^2, then, immediately, F must fall significantly as d becomes $d+\Delta d$ metres.

Immediately, the gravitational pull holding the satellite in its particular orbit would weaken, and the satellite would tend to start on its preferred tangential path. The whole process would rapidly and exponentially increase such that the orbit would be lost and the tangential path adopted.

However, this does not happen, and therefore, Newton's equation cannot be the right theory to use for orbiting satellites. The reason is that it works, and yet it should not because Newton uses an attraction force as his mechanism.

What happens, in reality, is that a launched satellite at any given orbital velocity will find its own, balanced, radius at which to orbit. In practice, of course, this is not left to chance, but is pre-calculated by scientists and the satellite is inserted into orbit at precisely the required angular velocity.

Absorption theory does not suffer from this Newtonian flaw. Because the satellite is, under absorption theory, independently generating its own gravitational acceleration towards the Earth, it does not experience a mutual attraction as proposed by Newton, nor does it have to respond to a distortion in spacetime created by the Earth, and which is, thus, connecting the two—our planet to the satellite.

Absorption theory provides the same concept of 'gravity wells' as does Einstein but calls them, more accurately, *potential gravity zones*. Therefore, it does not contradict the observed behaviour of satellites, nor does it contradict the law of angular momentum. In fact, the same explanation for the creation of *linear* inertia and momentum offered in this theory applies to the creation of *angular* momentum and inertia.

Einstein's spacetime wells have an exponentially-decreasing force gradient with distance from the planet or star concerned. Absorption theory matches that but explains how gravitational force is created, and the creation mechanism put forward permits matter to behave in the same way in response to gravitational force as does Einstein's. However, Einstein visualises a four-dimensional spacetime distortion field, whereas the author proposes a simple, more credible, *potential* gravitational field. In a way, this is similar to Einstein's spacetime in that his spacetime exists as a bland, neutral field, and it is not until matter is inserted into that field that a distorted curve is imposed and a 'gravity well' is created. It is deficient in its ability to satisfactorily explain *why* a gravity well is created and what happens to the gravity well inside a planetary or stellar object, or how that gravity well communicates its 'slope' or 'curvature' to the satellite concerned.

Absorption theory does not suffer from these limitations.

Absorption theory explains that, because the gravitational field's background wave flux is omnidirectional, as a satellite moves around its orbit it overtly behaves in the same way as for Einstein's spacetime theory. A satellite continues in orbit in accordance with the law of conservation of angular momentum, but the satellite in DOPA theory does not have to communicate with the Earth in the same way that Newton's and Einstein's satellite would have to.

According to DOPA theory, passing through the adjacent, oncoming point in space, to the satellite there is another net potential force that has the same strength as the one being abandoned, and that is also one that is pointed towards the centre of the planet. This one holds the satellite and turns its direction to keep it in the orbital path. This adjustment of position, keeping the satellite in a circular orbit, is conducted on a continuous basis and creates what is called the continuous radial centripetal force, which in the DOPA case, is the net induced drag towards the planet's centre.

In DOPA theory, the satellite passes from its weakening terrestrial link to the new, potential, omnidirectional field that will then continue to hold the satellite at its fixed gravimetric distance from the planet. Thus, the theory explains, without detraction, and without fallacy, how the differential potential gravity flux holds a satellite in a circular orbit. So, although DOPA theory also functions on an approximate inverse square theory, the difference between it and Newton is *the mechanism* that provides the stability. Newton's theory involves a minuscule information time lag between the centre of the planet and the centre of the satellite. This is what makes the incremental control theory for the orbit to be flawed. If Newtonian attraction were the actual mechanism, the time lag between the planet and the satellite would permit the delta movement of the satellite away along its tangentially-preferred course. In the case of DOPA theory, there is no time lag because the restraining centripetal force is created internally within the satellite's body, thus preventing any delta deviation of radius to occur. It is an essential difference, supporting Roberts' DOPA theory against both Newton's and Einstein's because both of those depend upon messages travelling between the objects along the fields linking them together. In Newton's case, it is a field of gravitational attraction; in Einstein's case, it is a field of warped spacetime, and the information as to how the warp is changing in the path of the satellite has to be (somehow) communicated to the satellite. Both are fields which the proponents claim have information flow along them at the speed of light. In the case of absorption theory, the information has flowed out from the planet long before the satellite encounters it in the form of a potential field that merely waits for the matter to enter it without any communication protocol.

Thus, when we are referring to an orbital situation. absorption theory does not allow the satellite to travel in incremental units along its orbit, it is held on a permanently-curving free-fall course and provides a logical, reasoned, theory in that context.

When considering elliptical orbits, in the same way as Einstein's gravity wells can accelerate and decelerate a satellite passing closer to and farther from a planet using an elliptical orbit, so potential gravity zones can impose the same effect. The difference is that Einstein's spacetime concept is not, itself, explained and has no stated and reasoned mechanism, whereas DOPA absorption theory does.

It is interesting that there are 'laws' such as Newton's and Einstein's and Kepler's, and yet these carry no explanation with them as to why or how they operate. They "just do". It is generally thought that, because something obeys a mathematical equation, it is sufficient unto itself to call it a law. And to the extent that it is useful, it is. But, that should not detract from the need to find an explanation for why these things happen the way that they do in the way that the 'law' describes.

A fundamental and critically-important difference between Newton, Einstein, and Roberts' theories is that in Newton and Einstein's theories, where a satellite planet orbits at great distances, such as the Earth from the Sun, there exists the problem of the mechanism whereby that planet reacts to the presence of its parent star. In Newton's case, he proposed an instantaneous and simultaneous link between every particle and every other particle in the universe. This implies an instantaneous, faster-than-light, method of communication which he was never able to explain and which is currently unacceptable, hence (through the arguments made above), Newton's attraction information must necessarily travel at the speed of light. Furthermore, although Einstein tried to accommodate his theory by saying that in his theory, gravitational effects (information) travel at the speed of light, he does not provide any explanation for what that effect is or how it operates. The difficulty with these two theories in the context of gravitational control can be exemplified by the theoretical question of what would happen to the Earth, under each system, if the Sun were to suddenly and instantaneously disappear. Firstly, under Newton's theory, how would the Earth's atoms, linked to the Sun's atoms, know that the Sun's atoms were no longer there? Secondly, the Sun's atoms would no longer be there, and if there were no atoms linked to those of the Earth, the further question arises as to what would be keeping the Earth in its circular orbit from the instant that the Sun disappeared. Since instantaneous information transmission is not acceptable, then one can only assume transmission at the speed of light, and so one is forced to assume that there would be an eight-minute time gap between the Sun disappearing and the Earth knowing anything about it. So, again, one is obliged to ask Newton what it is that is holding the Earth in its orbit for those eight minutes. There is no answer to that question. The same problem arises with Einstein's theory. According to spacetime theory, the Earth is orbiting inside a distortion in spacetime (or gravity well). And, according to Einstein's spacetime theory, the information along his spacetime field is transmitted at the speed of light. So, the same question must meet with the same silence in response. There is no answer to how Einstein's spacetime distortion can become aware that the Sun has disappeared until eight minutes have passed.

Now you can address the same question to the author. In this case, you will receive a meaningful and valid response. How would the Earth know, under DOPA theory, that the Sun had suddenly disappeared, and what would keep it in its orbit during the eight minutes needed for information to be received by the Earth? The answer is that, under DOPA theory, as has been explained earlier in this book, the Earth and Sun are not connected together by either attraction or spacetime. The Sun's gravitational zone through which the Earth orbits is a permanently-present potential gravitational field around the Sun. It comprises incoming pristine gravitational waves and outgoing degraded waves. The Earth tries to move towards the Sun in that centripetal gravitational field but is prevented from doing so by the centrifugal force of its orbit. At any moment in its orbit, the Earth moves and responds to a net gravitational force from the Sun that was created eight minutes previously. Thus, when the Sun disappears instantaneously, what happens is that pristine gravitational waves that would have penetrated the Sun now travel unobstructed towards the Earth, chasing the last of the depleted waves as they travel out towards the Earth. It will take eight minutes for the full-strength waves to reach the Earth, but in the meantime, the Earth is held in balance in its orbit by the spherical balance between the sunward pristine waves penetrating it and the Earthward degraded waves still travelling out from the Sun. During that time, this mechanism continues to hold the Earth in orbit. As soon as the new full-strength waves reach the Earth from where the Sun used to be, the Earth reverts to a balanced set of incoming pristine waves and no longer tries to drag itself towards the Sun. It is instantly released and continues on a tangential line off out into space exactly eight minutes after the Sun disappeared.

It must be recognised that, under DOPA theory, what is travelling out across space after the Sun has disappeared is not some mysterious force of attraction which would need a distant anchor, nor some mysterious distortion in spacetime which would need some energetic source of distortion, but a straightforward train of gravitational waves that is continuing to travel across space independently of the fact that the waves had passed through the Sun when it was there. These gravitational waves are entitled to be continuing through space under the laws of physics and are behaving properly in accordance with wave behaviour by continuing to travel towards the Earth at their maximum speed Cg. And when they have reached the Earth and have all passed by the Earth, the new, full-strength waves will reach the Earth following them in a normal fashion. There is no mystery involved. Gravitational waves do not need an anchor point; they travel through space under their own volition. They only perform a function when they enter into matter. DOPA theory is consistent and provides an answer to all such questions.

There is no other theory that can answer that question and explain that behaviour. Furthermore, and importantly, the answer to that question explains how Roberts' DOPA theory allows differential opposing partial absorption of gravitational waves to construct gravitational zones within the universal potential gravitational field that can control matter across the universe without any time lag and without paradoxes.

Matter tries to move according to the current state of wave amplitude differences and their net amplitudinal vector at any particular point in the universe.

The great benefit of absorption theory is that it explains a mechanism that accounts for the creation of gravitational force, the way that matter tries to move, and explains the existence of inertia, and momentum, thus accounting for why various eminent laws apply and operate.

It also by-passes and displaces Van Flandern's arguments (in a 1998 paper concerning Hendrick Lorenz's version of relativity) relating to the need for gravitational waves to exceed the speed of light. He argues things such as the speed of gravity is far faster than the speed of light—in fact, of infinite speed. Roberts considers this to be preposterous, just as it was for Newton to imply it. However, like Newton, he could not see any way past it.

Fortunately, absorption theory does not need such hypothesising.

Van Flandern also made such statements as, *"If gravity propagated between the Sun and the Earth at the same speed as visible light, the Earth would..."* He saw the force of gravity as being, as Newton did, some kind of thing that propagates between the Sun and the Earth in equal amounts. This is unsupportable hypothesising, and he has made no progress, learned no lesson from Newton's same mistake. As described above, absorption theory accounts for why the force generated within disparate objects is equal (See figure 35). This is the shadow-generated force that is exerted within both objects simultaneously to make them want to move towards each other.

Van Flandern, Newton, and Einstein fall into the same 'time trap'. For example, they each consider that, somehow, the gravitational force emanates from the Sun and exerts a force on the Earth. In doing so, they necessarily have to wonder at the eight-minute time lag between the Sun and the Earth and wonder how it can operate. Absorption theory avoids such dilemmas because the accelerating drag force that it experiences is created locally within the Earth as a result of the pre-existing potential field through which it passes. In simple terms, the potential gravitational field is already in existence, and the planet Earth merely passes through it, responding to it as a consequence.

This explanation for the continued control of objects at vast distances is so fundamentally important to absorption theory that, recognising its complexity, the author makes the exception of choosing to address this matter a second time and in different words, hoping that it will be more-clearly conveyed by that means. So here is a repeat of the argument:

First, consider light. It is commonly accepted that it takes about eight minutes for light to travel from the Sun to the Earth. Consequently, people accept the proposition that if the Sun were to, somehow, disappear instantly, we would continue to see it for a further eight minutes. That seems both apparent and logical. The light is, in fact, the information that we rely upon to tell us that the Sun is still there and yet, because of the time lag, we are deceived.

Now consider force. If there is as an attraction force between the Sun and the Earth, and the Sun were to disappear suddenly, then what would happen? In this case, matters are not apparent or logical. The same applies to both Newton's attraction theory and spacetime warp connection.

Using the light analogy, it would take eight minutes for the Earth to 'realise' that the Sun was no longer there, so, in the interim, what course does the Earth continue to take? Does it continue along its circular orbit, which would imply that the 'force' between the Sun and the Earth still exists even after the Sun has gone? That is much more contentious than the idea of light simply stopping eight minutes later.

At the instant when the Sun disappears, the Earth cannot 'know' that the Sun has gone, because the relevant information would take eight minutes for it to arrive at the Earth. But, the paradox is, *how can the Earth continue in a circular orbit when the Sun is no longer there?* This paradox attacks the root of any direct connection theory or any space distortion theory. DOPA theory is the only one ever proposed that avoids this paradox **because it provides the mechanism for the Earth to continue in its circular orbit for eight minutes until the information arrives that the Sun has gone. That is unique to this theory and fundamentally important!**

In fact, the Earth is continually being controlled in its solar orbit by information that is eight minutes out of date. The potential, spherical gravitational zone that the Earth occupies was created eight minutes previously by the Sun's partial absorption of pristine gravitational waves. This continuous field-creation effect is unnoticeable because, until matter is placed within the zone, the field does no work. Similarly, we accept that we cannot see the destruction of an exploding star until several million years after the event, and so we must accept that the gravitational field created between the Earth and that star was created several million years ago. We do not notice any effect, because it is so incredibly weak as to be non-existent. And yet, absorption theory accounts for this at the million-year scale in exactly the same way that it accounts for the Earth's behaviour in the event of the Sun disappearing.

And absorption theory has the clarity of stating that *gravity itself has no velocity*; gravitational waves move at the speed of light, but gravity is simply the force generated within matter. It does not, of itself, move at any vector velocity.

DOPA theory says that gravitational waves have already been travelling up to the moment of utilisation from all over the universe and are not specifically created as a mysterious force by the Sun or the Earth; they are already there, everywhere, universally present and omnidirectional, and when they penetrate both the Sun and the Earth, the spherical 'shadow' field is then generated. The Earth simply sails through the Sun's potential gravitational zone and tries to drag itself towards the Sun, and that drag is resisted by centrifugal force. The Sun and Earth, in a way, have nothing to do with it at that instant of time. They have already made the potential field and its gravitational shadow between them eight minutes previously—it exists between them permanently but is not a connection; it is just a zone of depleted gravitational waves.

Mentioning another famous name, Van Flandern claimed that Lorentz's version of relativity could easily account for accepted observations if one simply assumes a local gravity field. But neither Lorenz nor Van Flandern (in fact, no one) has been able to provide a sensible explanation for such a phenomenon until now. DOPA theory does just that—it is both locally-created and much more locally terminating than infinity. Thus, absorption theory provides the mechanism that, like Einstein, Hendrick Lorentz was seeking so long ago and never found. This example highlights the problems associated with just making a statement and hypothesising without any explanatory mechanism. Many people have done this and it is wasteful of effort because without a mechanism it is just a dream.

The propositions described in this Chapter of the book are all new concepts produced by this theory and reveal significant differences when compared to any previous works. But more than anything, they all carry logical explanations based on practical, effective working mechanisms.

4.6 How does the theory explain phenomena such as elliptical, asymmetrical, and other rogue orbits such as Mercury's?

The theory does not explain and does not need to explain such phenomena as it is not a theory about how matter behaves in response to gravitational force.

Nor does it purport to replace either Newtonian or Einsteinian mechanics. DOPA theory describes the creation of gravitational force and constructs gravitational field structures *equivalent to* the relativistic spacetime gravity well mesh—it provides a reasonable basis for their existence as opposed to saying, as both Newton and Einstein did, that one has to just accept them as existing because they say so.

4.7 How does the theory explain why light is bent when it passes close to the Sun and other dense cosmic bodies?

The theory does not explain and does not need to explain such phenomena as it is not a theory about how matter behaves in response to gravitational force. This theory only offers an explanation as to how gravitational force is created, not how matter behaves in response to gravity.

Having reiterated that point several times, some observations can, nevertheless, be put forward and comments can be made on possibilities. However, please note clearly that Roberts' theory is not involved in theories of how light behaves close to stars.

The phenomenon is well known and recorded. It is known as 'lensing' when, passing a large star or black hole, light is observed to be bent in its path (Figure 53.)

Relativitists say that only relativity can account for this phenomenon. However, that is not necessarily correct, although if it turns out to be correct, that does not affect Roberts' theory. This is only emphasised because alternative possibilities are discussed below.

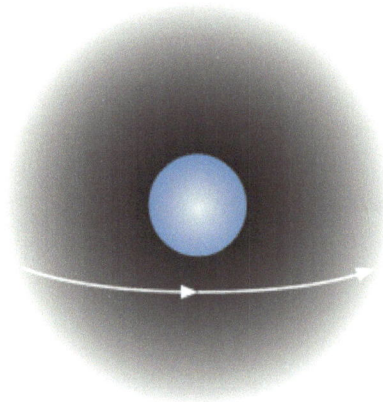

Fig 53. Light lensing around a star

For example, two sets of laboratory research work completed in 2011 and 2015 by Rancourt (and Tattersall) demonstrate that coherent light can affect gravity by reducing its strength when placed in its path. If that research work were to be validated, then that alone would provide a reason just on the basis of opposite actions and reactions—if light can control gravity, then gravity can control light. Also, of course, the Sun is an extremely intense source of light at close quarters.

[Louis Rancourt, "The Effect of Light on Gravitation Attraction", Physics Essays, 24(4), 557-561 (2011)]

[Louis Rancourt & P.J. Tattersall, "Further Experiments Demonstrating the Effect of Light on Gravitation" Applied Physics Research Vol 7, No 4 (2015)]

As a second potential concept, all around any star, including our Sun, there is an intense *potential gravity zone*. Every atom or speck of matter ejected from a star's surface has to pass through the potential field (its own *potential gravity zone*) and, while within it, in addition to magnetic pull, the differential gravity wave field exerts its net force inwards, back towards the star's centre.

It is not difficult to see that, if every atom and speck of matter is affected or controlled by the potential force field, then it will be either ejected violently, dragged directly back into the star, as shown in Figure 50, or will become part of a zone of orbiting matter that resists being accelerated inwards by virtue of its high orbiting velocity. Ultimately, all non-returning particles will be part of a vast solar atmosphere that ultimately extends out for millions of kilometres from the Sun in the form of the solar wind.

The solar atmosphere will, it is reasonable to propose, be denser towards the Sun's surface in the same way as the Earth's atmosphere is increasingly dense towards its surface.

The author, therefore, offers another possible explanation that is unrelated to this theory. Perhaps, lensing is caused by refraction as light passes through the lens of changing density and cooler-temperature gas surrounding the Sun or any other star. Stars and black holes alike have zones of particles adjacent to them which could act as density lenses. Consider, for example, the well-known phenomenon of mirages observed by eye or camera in hot countries where the ground surface becomes heated to a higher temperature than the air above it. In turn, the hot ground surface heats a relatively shallow zone of air above it which cools with height. The physical result is that there is a temperature inversion established contrary to the usual state of warmer air rising above cooler air below. In the case of this phenomenon, light becomes bent to give optical illusions of the presence of water on the ground—mirages. We are familiar with the idea that *light is bent towards denser air,* as shown in the following diagrams:

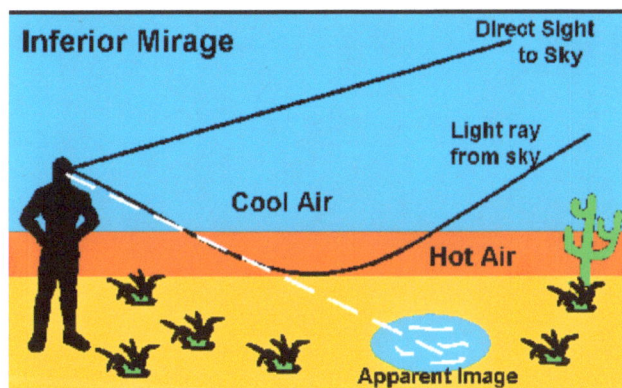

Fig 54. Refraction of light towards denser air.

Fig 55. **Light refracted from the blue sky looks like patches of water on the sand, even though there is no water present.**

Fig 56. Light refracted from the blue sky looks like water on the road, even though there is no water present.

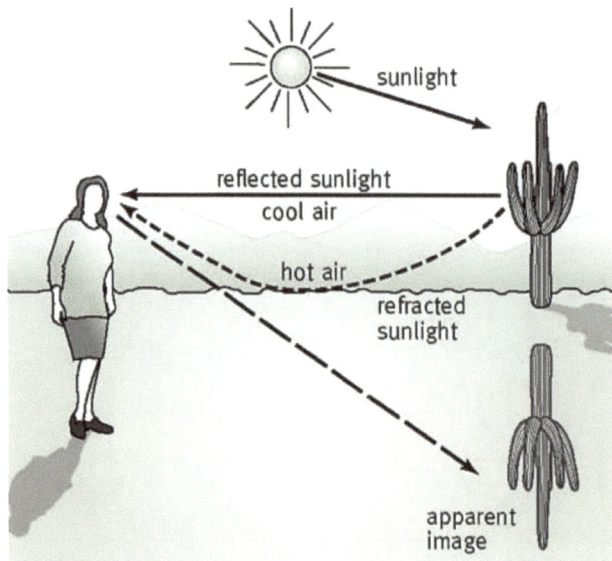

Fig 57. Refracted light creates inverted images as if in a mirror when it is bent towards denser air.

Fig 58. Refraction of light causes inverted images that look like reflections in water.

This same, simple, principle can be applied to the situation where light is bent towards cooler denser air adjacent to the planet's surface, as shown in Figure 59. In this case, an observer standing on the shore can see the inverted shape of a yacht which is over the horizon and which should be concealed.

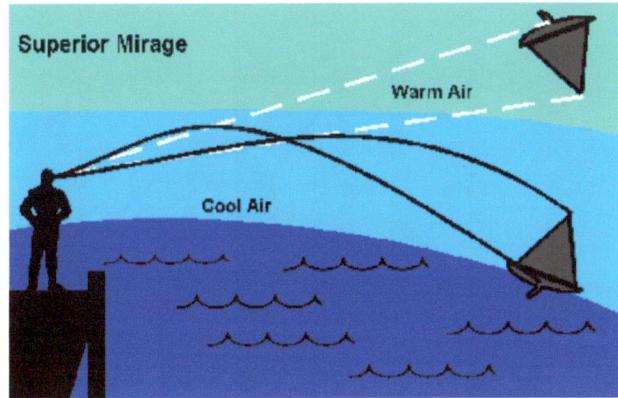

Fig 59. Light refracted downward allows an observer to see a yacht that is over the horizon and should be concealed.

This diagram shows the same scenario as an observer on Earth seeing a star that is behind the sun and which should be concealed, as shown in Figure 60.

Currently, it is considered that somehow, through the unexplained agency of relativity theory, light passing from a star behind the Sun is bent even though the light has no mass, such that an observer on Earth can see the star as a 'mirage'. By "unexplained," the author does not mean that there is no mathematics to forecast it, and does not imply that the forecasts have not been ratified, but considers that, because the basis of the mathematics is spacetime theory, which has no evidence to support its existence nor supportive evidence for its functioning and behaviour, the causes of lensing could be based on the wrong mechanism.

What the author has never read is whether or not lensing produces inverted images. We would not be able to discern inversion on a round dot of light such as a distant star, but we might be able to observe it in a close group of stars whose relationship we know from when the Sun is not being eclipsed. If it were to be so observed, then that might support the idea of physical density layered refraction. At least this is another possibility for research.

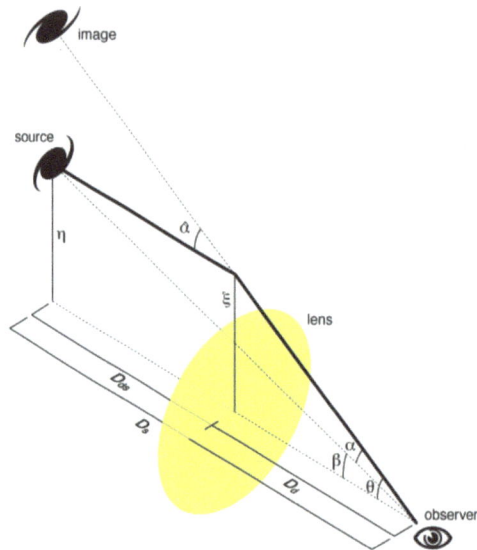

Fig 60. Light bent around a star is reputed to be caused by relativistic physics, but might be caused by refractive lensing owing to density gradients within the star's immediately-adjacent heliosphere.

As stated earlier, this discussion is not a direct part of absorption theory, because if a physical process occurs, then the explanation of how gravitational force is created does not affect how that gravitational force affects cosmic bodies. The author has expanded this question just for the sake of academic interest. However, since DOPA theory challenges the very existence of spacetime, it may be reasonable to suggest that looking elsewhere into relativity or elsewhere, may be necessary to explain stellar lensing of light.

4.8   Does the theory conflict with why the force of gravity would appear, according to theoretical calculation, to increase towards the centre of the Earth until the outer face of the outer core is reached?

It is interesting that whilst absorption theory can account for shell theory it disagrees with the PREM theoretical proposal that gravitational force initially increases with depth in the case of a densely-cored star or planet such as the Earth. This subject has also been discussed in detail in the vicinity of Figure 25 above.

In the first instance, the current PREM calculations that indicate a likely increase of gravitational force appear to be based on Newton's theory of attraction in order to produce Figure 61. The author has checked this using Newton's equation, and it seems to be so.

Newton's equation produces this false effect because of the inverse square of the distance incorporated into the equation, and because his shell theory, which relies on attraction as being the force of gravity, being a discredited mechanism. Relying on these two items, gravity would unjustifiably increase as shown in of the upper red graph in Figure 62, which is based on the use of Newton's equation and approximates the PREM's dark blue curve.

74

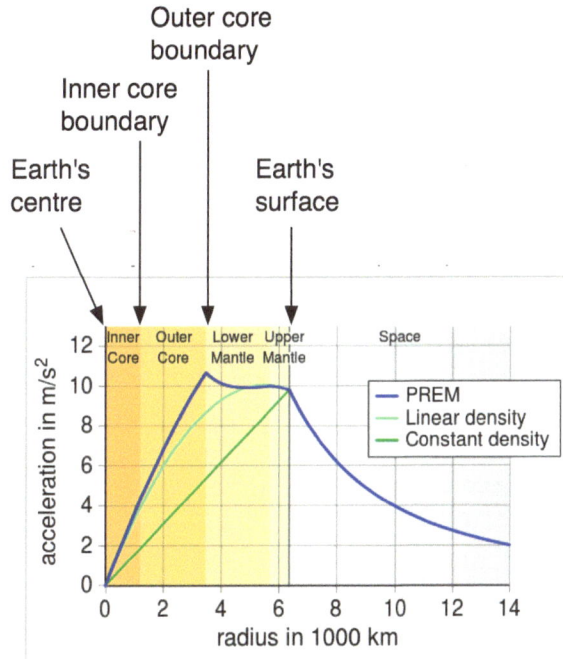

Fig 61. The blue line shows gravitational acceleration to increase with depth to a maximum at the outer core boundary. According to the Preliminary Reference Earth Model.

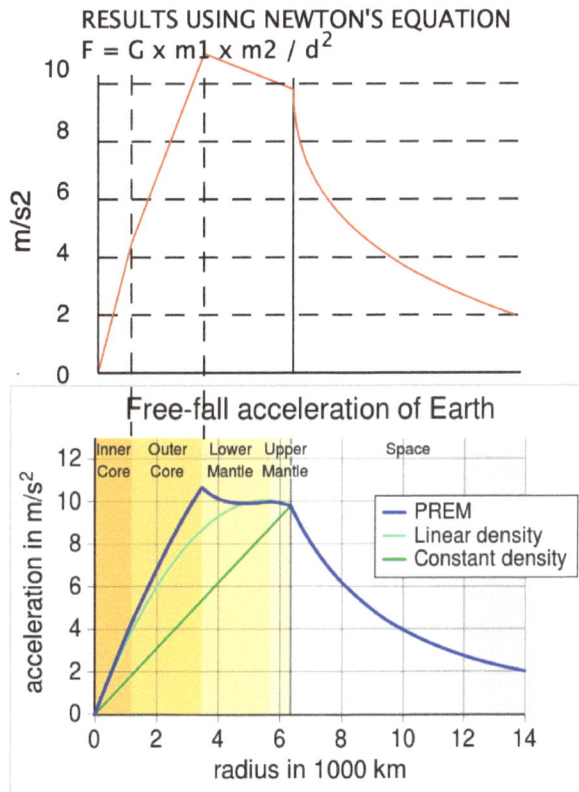

Fig 62. Earth's gravitational acceleration externally and within the planet (upper graph plotted red according to Newton's equation). Lower graphs are the Preliminary Reference Earth Model.

On the contrary, based on his absorption theory, and based on multiple-scenario assessments, the author concludes that net gravitational force will decrease inwards, past any cores present, falling to zero at the centre of the planet. The form of its plot will loosely resemble the light green curve in the PREM plots.

Why is that?

Firstly, we must recognise that since Newtonian gravitational attraction is no longer considered to be a valid concept, then it is invalid to use it to calculate the gravitational force with depth by discounting the effect of any shell. The important point to recognise here is that absorption theory produces the same effect as Newton's law outside a body that is generating its own potential gravitational field *but not within*. Inside a planet or star, only absorption theory provides a real, graphically-demonstrable solution.

What does Newtonian shell theory say happened within the Earth below its surface? It says that, as we go down into the Earth towards the outer core, all gravitational force exerted by the continually-thickening shell above us becomes a net zero. Shell theory states that if the body is a spherically symmetric shell (i.e., a hollow ball), no net gravitational force is exerted by the shell on any object inside, regardless of the object's location inside the void within the solid shell. In other words, if the Earth were a hollow shell only a thousand kilometres thick, then within it, one could float anywhere under conditions of zero net gravity. DOPA theory agrees with the net zero-gravity shell outcome but provides a better basis for shell theory gravitational force calculations than Newton's inverse square law. (see the text around Figures 25 and 26 as well as here.)

In both Newton and Roberts' theory, one would not be in a state of zero gravitational force. Instead, there would be a state of *net* zero gravitational force owing to all forces cancelling one another out anywhere within the internal void. Surprising, but both theories support this.

It might be helpful for some readers, to explain, at this point, a simple principle of planetary surface gravity that is also compatible with absorption theory. The force of gravity experienced at the surface of a planet or star relates to both the mass of the planet and its radius. This phenomenon was first expressed by Newton when he deduced that the force of gravity expressed from the centre of a cosmic body varied inversely as the square of the distance from its centre.

Let us consider three planets of varying density and radii in order to calculate the surface gravitational force using Newton's theory.

Planet A. Density 5 g/cc and radius 5,000 km.   Mass = 2.62E+24 kg   g = 6.99 m/sec^2
Planet B. Density 5 g/cc and radius 1,000 km.   Mass = 2.09E+22 kg   g = 1.39 m/sec^2
Planet C. Density 625 g/cc and radius 1,000 km. Mass = 2.62E+24 kg   g = 174.2 m/sec^2

These three sets of figures say all that is needed. Consider planet A, having just a little smaller radius than Earth, with an average density of 5 g/cc (5,000 kg/m^3) similar to Earth. It produces a surface acceleration of 6.99 m/sec^2. Now, if we reduce its radius to planet B size, but keep its density the same, we find that its total mass is reduced by about 100 times and yet its surface gravitational acceleration only falls by about one third to 1.39 m/sec^2. That is because of the effect of the inverse square denominator in Newton's equation. Now, look at planet C, which is planet A shrunk down but keeping its original mass. If it were just a matter of mass, one would think that the surface acceleration would be the same, but because of the inverse square denominator, its gravitational force has jumped up to 174.2 m/sec^2. So, this highlights the effect of reduced radius changing with constant mass.

Finally, it is necessary to know how absorption theory creates the same effect. The answer may be assessed from an examination of Figure 35. Notice how the radius of the smaller planet is about half of that of the larger one. Also, the wave absorption is about 25% pristine as opposed to about 55% for the larger planet. Bearing in mind that the acceleration experienced under absorption theory is the difference between pristine and the absorption, then, given the same density in both planets there is a reduction in volume to much less than a half, but an absorption that is 'linearly' only about half. We can see the similarity here. If the density of the smaller planet were to be about one hundred times bigger than the larger one, it can be seen that the increase in absorption would be about two hundred times that of the larger, less-dense planet.

Pristine is 100 m/sec^2. The density of both planets is the same.
Absorption in large planet is 55 m/sec^2, so accel = 100 - 45 = 55 m/sec^2.
Absorption in small planet is 25 m/sec^2, so accel = 100 - 75 = 25 m/sec^2.

Thus, we can see that the absorption theory matches Newton's in principle in that the smaller planet had a decrease in gravity of about a half of the larger one, similar to that when Newton's equation was used. i.e. a lesser fall than could be accounted for by the much greater reduction in volume, as shown between planets A and B above.

4.9   How does the theory explain why heavy metals and minerals sink preferentially downward within the molten planet to form the core?

By the standard hydrostatic principles of buoyancy.

The DOPA gravitational wave principle that applies to an atom in space, or to an entire planet, also operates on any small element of denser metal within the molten magma below the planet's surface. The mechanism is precisely the same as that illustrated in Figure 40 and its subsequent text.

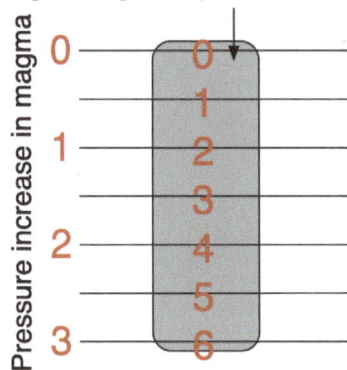

High density metal crystal or globule
of S.G. = 6 measuring 1 metre tall
floating in magma liquid of S.G. = 3

Pressure increase in magma

Downwards pressure increases by 6
units in the crystal, but only 3 units in
the surrounding magma, resulting in
a net excess pressure down at the
base of the crystal of 3 units/m^2

Fig 63. A high-density crystal of specific gravity 6 g/cc (or metal globule) floating in a lower density magma of specific gravity 3 g/cc.

A globule of dense metal or a denser crystal formed within a magma chamber at depth within the crust will exhibit a tendency to move downward preferentially because, although the acceleration of gravity is the same for both the globule/crystal and the surrounding magma, means that it absorbs more energy from the net, downward, incoming, waves and thus experiences a greater net downward internal drag than the surrounding liquid magma in which it is suspended. Figure 63 illustrates the principles of wave absorption that Roberts theory provides and which conform to the principles of standard hydrostatics.

Consider, as shown in that figure, a crystal or globule of dense metal floating in a magma chamber within the Earth. The net DOPA gravitational force is downward towards the centre of the planet, and it is directly proportional to the density of the material through which the gravitational waves are passing. In the magma, being less dense, only 3 units of increased pressure are experienced immediately down the outside length of the crystal. Within the crystal, however, having a density twice that of the magma, the net absorption of gravitational waves has produced a downward pressure double that of the magma. Thus, at the base, there is an upward buoyancy force of x+3 units per sq.m. and within the crystal or globule, at its base there is a downward drag force of x+6 units, leaving a net downward force that drives the crystal towards the centre of the Earth. By this means, the gravitational formation of a dense core follows automatically. Also, in the Earth's case, there has been the formation of an extremely dense inner supposedly-solid core and a less-dense outer liquid core, both of which have been created by the same mechanism.

A very interesting corollary to this concept is that, in addition to the differential mechanism above, the fact that there is an upward-travelling wave through the crystal means that the extra-depleted wave passes out from the crystal, upward into the magma above. This process imparts a minuscule increased downward drag to the liquid immediately above the crystal relative to the surrounding

78

liquid as if its density were effectively increased. That provides a mechanism for the overlying higher liquid to move down and not only follow the crystal downward but to place an additional minute downward pressure onto the top of the crystal, thus providing a mechanism to help overcome the liquid's viscosity. At the same time, the preferential absorption of the downward waves reduces the net downward drag in the magma immediately below the crystal. These differentials will be minute and will dissipate rapidly within the almost-incompressible fluid magma.

Nonetheless, this feature provides an interesting and testable potential for experimentation and computer analysis which could confirm the partial absorption concept of the Roberts theory. When sufficiently sensitive instruments can be developed to test this is not known by the author. Indeed, it may be possible that such instruments already exist.

Concerning the crystal example, inspection shows that absorption theory supports the standard principles and mechanisms of hydrostatics and buoyancy.

4.10 How does the theory fit in with the concept of an expanding universe whose edges appear to be expanding outward at an accelerating rate?

The reader may not be too sure about the term 'expanding universe'. In current astrophysics, it is considered that it is the very fabric of the universe itself that is expanding and that matter within the universe is being carried outward within that expanding universe passively; it is not the matter that is expanding through a static universe.

The absorption mechanism for the creation of gravitational force does not fit in with the current concept of an expanding universe. Instead, it replaces the concept of the universe itself expanding with that of the sphere of matter expanding within a much larger or infinite universe. DOPA theory provides the necessary mechanism, explaining why gravitational force drives peripheral matter radially outward at an increasing rate. Again, there is no proposed mechanism for the fabric of the universe expanding, but there is a proposed mechanism for Roberts' proposition. That mechanism is the same differential opposing partial absorption theory.

It is one of the tenets on which this theory is founded that the omnidirectional gravitational wave flux that we experience in our location of the universe may not only be uniform here but may, logically, remain uniform spatially over the great majority of the universe. However, in addition to the 'big bang' gravitational waves, since such material activities as the merging of neutron stars and/or black holes cause bursts of gravitational waves that will undoubtedly spread outward, we can, without any difficulty, imagine that the conditions pertaining in the peripheral parts of the universe are markedly different from those at its centre.

The further towards the centre one is, the more uniform gravitational wave conditions can be expected to be in terms of both direction and amplitude. However, at the periphery, it is inescapable logic that all gravity waves nearer the outer periphery become increasingly mono-directional in a radially outward direction the closer to the periphery they are. This is particularly so if the universal gravitational wave flux has been produced primarily by the activities of rotating binary star couples.

The outcome of this logical deduction, based on the mechanical presumptions of the Roberts theory, is that the generation of strong radial drag forces within matter at the periphery of the universe will increasingly accelerate matter outward at a further-accelerating rate in just the way we believe happens from our current scientific observations.

Ultimately, with no equal-and-opposite gravitational waves coming from outside the material periphery, peripheral matter will become subject to the full, unadulterated, gravitational wave drag force of pristine waves as gravitational waves pass through it at the in-vacuo speed Cg, being the same as the maximum speed of light C.

Figure 64 has been included at this point as a matter of historical interest. It is taken from Roberts' original paper of 1978 which presumed that gravity exerted a pure repulsion force. This theory was not acceptable at the time because, although the concept of wave penetration and absorption were foreseen, the critical consequences of opposing waves were not. Hence, the initial theory foundered. It is, for historical interest, included at Appendix 1 for study.

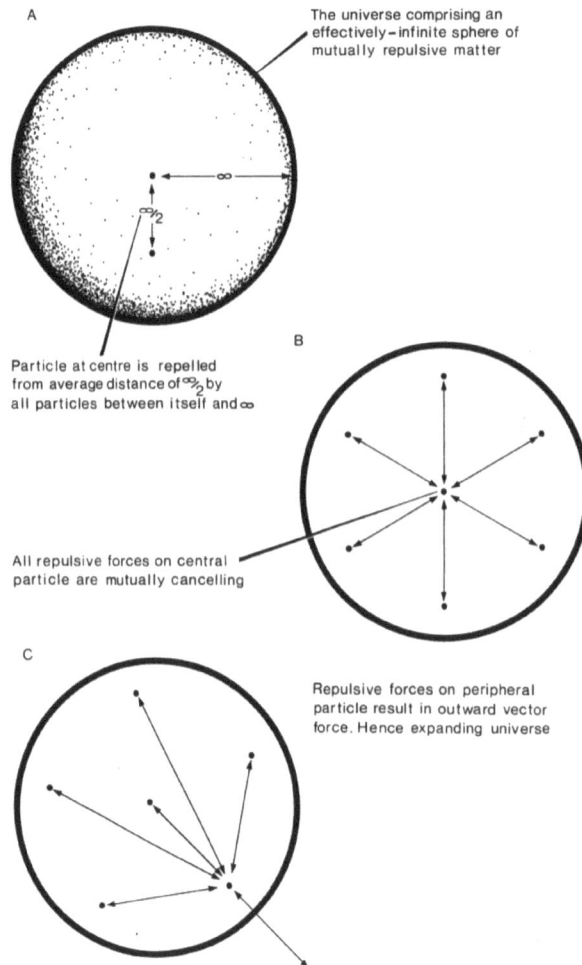

Fig 64. Roberts' original, 1978, concept of matter expanding within the universe by virtue of a pure repulsion force. Now discarded.

4.11 Does the theory fit in with what we know about how stars develop?

In the simplest terms, the answer to this question is 'yes'. This theory provides the mechanism for how galactic dust clouds condense and coalesce into stars with or without associated planetary satellites.

Beyond that, there are no apparent conflicts caused by this theory in connection with the various stages that stars pass through during their evolution. The reason is that this theory does not relate to the way that cosmic bodies behave in connection with gravitational force, it merely provides the mechanism whereby that force is created. Nonetheless, question 4.12 contains some particularly relevant input on this topic.

4.12 How does this theory's mechanism fit in with the formation and development of black shell stars (inappropriately known as black holes)?

It is proposed that absorption theory provides an improved concept for the formation of black shell stars than that provided by Schwarzschild and his misnamed, misplaced, single, thin, 'event horizon' and its associated singularity.

Light Phase 1. Leading up to becoming a black shell star.

Figure 65 shows a proposed approximate structure of a large star before acquiring sufficient mass to enter a black shell stage.

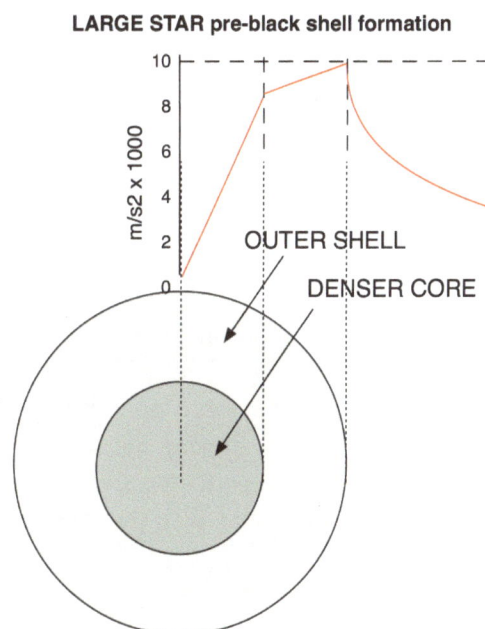

Fig 65. A large star with dense core prior to entering the black shell (black hole) stage Light Phase 1.

81

Note that the net gravitational force is zero at the centre of the star, increasing rapidly, as might be expected, outward through the core and less rapidly from the core towards the star's surface.

At this stage, peak gravitational force (or acceleration) is close to the surface of the star and almost equals the speed of light.

Light Phase 2. Development of the star after initial formation of the black shell.

Figure 66 shows the star at the Light Phase 2 point where it has acquired sufficient mass that its peak gravitational pull now matches or just exceeds the ballistic escape velocity of light. In different stars, the rate of reaching this stage would be faster or slower. The primary control mechanism for this will be the rate of acquisition of mass by the star. If, for example, a star is one of a binary pair and is feeding off its partner, then its rate of mass acquisition will be very high. In such a case, the uniform surface shell will be developed quickly.

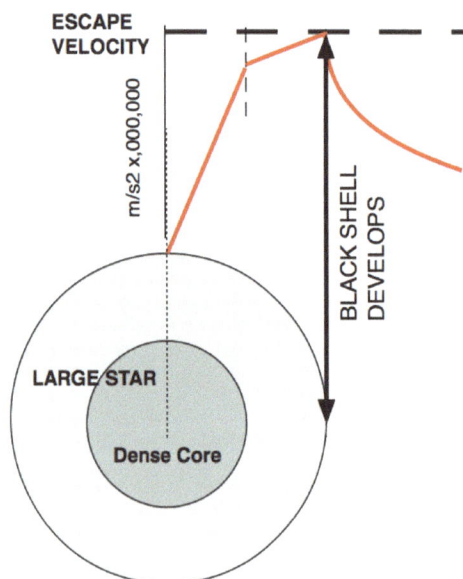

Fig 66. A very large star with dense core having just reached the initial black shell stage Light Phase 2.

Light is not only created at a star's surface but at all levels. Light created deep within a star can take tens of thousands of years to reach the surface from where, ultimately, it is emitted. So, at this initial black shell stage, all light reaching the surface or emitted directly at the surface will be prevented from leaving the surface, which will appear black from the outside. This provides a source of speculation. For example, what happens to the light that is not permitted free passage out? Does it accumulate just below the black shell? Will the star become brighter within and start to increase in temperature?

In Light Phase 2, any radiation travelling tangentially above the star's surface will be diverted by the gravitational field but will not necessarily be captured by the star. However, any radiation or matter that naturally travels towards the star's surface will be captured and held below the surface.

Owing to turbulence at the star's surface, it is highly likely that the formation of the black shell will be intermittent and will manifest itself as black patches forming and dissipating at the surface over time. How long it will take for the black patches to become contiguous is unknown, but, by human standards, it is likely to be a very long time. Eventually, however, a continuous thin black shell will establish itself permanently at the surface of the star.

To divert from the main thread of the arguments for a moment, it is interesting to read this scientific news item:

"Early last year a far-off star captured the imagination of the scientific community thanks to its incredibly bizarre behaviour. The star, which is officially named KIC 8462852 but is better known as 'Tabby's Star', frequently goes dim at unpredictable intervals, and researchers were initially at a loss as to how to explain it. later research suggested a cloud of dust was the culprit, but earlier this month the star's light began to wane once more, and this time it got darker than ever."

Perhaps this is the kind of stellar behaviour that we could expect from a star at the point of becoming a black shell star? Once the author's concepts are known, this could open up opportunities for observers to look out for patchy behaviour in luminosity.

Interestingly, Phase 2 is not the end of the shell star's evolution.

Light Phase 3. Development of the shell into a thick zone with an inner and outer surface.

There is no reason why the black shell star should not continue to increase in mass, and this process increases its maximum gravitational force beyond the escape velocity of light and other EMF radiation.

Because of the increase in total gravitational attraction, the original black shell widens as shown in Figure 67. In that figure, the red line shows the Phase 2 shell configuration whilst the green line shows the Phase 3 wider shell generated by the increased gravitational field strength.

In Light Phase 3, because the gravitational attraction of the star is greater than EMF escape velocity at any level *within the wide shell,* any light created *below the inner surface of the shell* is drawn to and held at the inner surface of the wide shell and accumulates at that horizon. Any light created *within the width of the shell* within the star is attracted inwards towards the inner shell surface. Thus, all light and any other EMF radiation continue to be collected at that inner surface horizon. The inner black shell surface thus becomes a kind of 'thin shell' in its own right, comprised of a zone of accumulated EMF radiation that can go nowhere. EMF radiation is dragged to that level from both above and below. (See Figure 69.)

In Light Phase 3, the maximum gravitational force is developed at the surface of the star, but because it is so much greater than escape velocity, the outer surface of the shell now exists above the surface of the star somewhere in space as shown by the green line in Figure 67. This will be anywhere between the surface and a great distance out depending on the developed width of the black shell.

The concept of a wide black shell having inner and outer surfaces is new, being a direct, logical, consequence of the DOPA absorption theory.

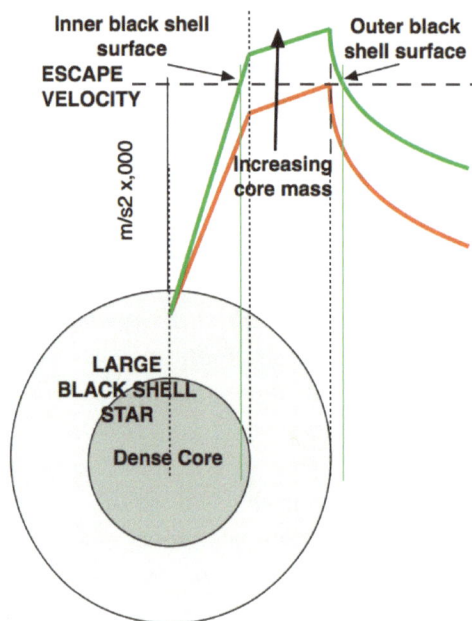

Fig 67. A very large star with dense core having developed a wide black shell Light Phase 3. (Coloured green.)

The outer shell surface, now situated a short distance above the star's surface in space, is the equivalent of what is currently called the 'event horizon' because even tangential radiation cannot escape from beneath it. In the context of absorption theory, it is called the "outer black shell surface", as shown in Figure 67. Above the star's surface, but below the outer black shell surface, any radiation travelling tangentially between the star's surface and the outer surface of the black shell will necessarily be captured by the star. (See Figure 70.)

Above the outer shell surface, any radiation travelling tangentially will be diverted by the gravitational field but will not necessarily be captured by the star.

However, any radiation or matter that falls naturally towards the star's surface will be captured and held below the surface, and in the case of light and other EMF radiation, it will then be dragged down to just above the inner shell surface.

The absorption mechanism in DOPA theory does not utilise an inverse square law within the mass of a body itself but employs an approximate one outside the body of the star or planet concerned.

In Newtonian, Einsteinian and Robertsian principles it is recognised that the greater the radius of a spherical body of a given mass, the less will be its surface gravity. The converse, of course, applies. That is because of the inverse square law in Newton's equations and the geometry involved in absorption theory—combined with the concept that gravity functions as if it acts from a point at the physical centre of the body.

The problem that this interpretation brings with it is the counter-intuitive one that, as a star becomes more massive and of greater radius by virtue of acquisition of mass, its gravitational surface acceleration could actually decrease. Therefore, in summary, the gravitational force exhibited at the surface of a cosmic body is a function of both mass and radius.

These principles encouraged the creation of Schwarzschild's equations that necessarily led to the automatic conclusion that once the compaction of a star's core starts, its self-gravity increases exponentially, leading it to collapse under its own gravitational stress to become a 'singularity', which is a non-dimensional point in space that contains all of the mass of the original star's core. It is a mind-blowing, theoretical, concept that the author is unable to comprehend, let alone agree with because it is nothing more than a mathematical construct.

However, as part of the development of DOPA theory, a new idea came to light that can explain why any Schwarzschild singularity may be prevented from developing. The development of gravity-free, pressure-limited, cores to stars is discussed below in Section 5.5. **The consequence of this is that there is no need to invoke Schwarzschild's' star-collapse singularity in order to generate the development of a black shell star.**

Newton and Einstein's theories were primarily developed in relation to bodies moving in space. It is, in the author's view, a false premise for either Newton or Einstein's work to be extrapolated to include its behaviour within a body. It is only absorption theory that stands up to inspection within a body—and that is because it is a theory specifically developed in relation to the behaviour of gravitational forces within a body, where they are created, *as well as* away from a body's surface, such as in space, where they create gravitational zones and the universal gravitational field.

As a result, the following calculation shows that, following Roberts' concepts, an exceedingly large star—even, as an extreme case, with a very low internal average density of only 13 g/cc—can develop a surface escape velocity equal to that of light and therefore transmute undramatically into a 'black hole'. The conventional Newtonian equation for determining the ballistic escape velocity from a spherical body is:

$$v = \sqrt{\frac{2GM}{r}}$$

Equation 4.1 The Newtonian equation for calculating the ballistic escape velocity from any spherical object whose mass is M kg. v is in metres/sec, G is Newton's gravitational constant, and r is the object's radius in metres. v is independent of the mass of the escaping object.

Using equation 4.1, Table 4.1 has been produced as the outcome of a calculation for a star which increases in mass and consequently expands to the point where its radius is the same as the orbital radius of Venus, 108 million kilometres, and having an average density of only 13 g/cc. (Which is chosen impossibly low just for numerical argument to make the point.)

G	6.674E-11	N(m/kg)2
r	108,000,000,000	m
r	108,000,000	km
D	13.00	Average
Volume	5.28E+33	Cu.m. Vo
Mass	6.86E+37	kg
v	291,170,810	m/sec
v	291,170.81	km/sec
v	1,048,214,918	km/hr
v	651,065,166	miles/hr
C	299,792,458	m/sec
	299,792	km/sec
	1,079,252,849	km/hr
	670,343,384	miles/hr

Table 4.1 The size of a star needed to become a black hole with an average density of only 13 grammes/cc is the radius of the orbit of Venus—108 million km.

If we increase the average density of that star to the same as our own Sun's core (150 g/cc), which is, clearly, not an outlandish suggestion, then, to turn into a black shell star its radius would have to be only about 35 million kilometres—less than half of the orbit radius of Mercury! By galactic standards, little indeed. See Table 4.2.

insert numbers in red		OTHER PLANET OR STAR		
mass of spaceship (kg)	m	3,000	kg	irrelevant
	tonnes	3	tonnes	irrelevant
Gravitational Constant	G	6.674E-11	N(m/kg)2	
Radius of planet	r	35,000,000,000	m	
	r	35,000,000	km	
	D	150.00	Average density of planet (5.49 Earth average)	
	Volume	1.80E+32	Cu.m. Volume of planet	
	Mass	2.69E+37	kg	<<< Checked OK
Equation for escape velocity	v	320,528,077	m/sec	
	v	320,528	km/sec	
	v	1,153,901,076	km/hr	
	v	716,708,743	miles/hr	
speed of light	C	299,792,458	m/sec	FIXED
		299,792	km/sec	
		1,079,252,849	km/hr	
		670,343,384	miles/hr	

Table 4.2 The size of a star needed to become a black hole with an average density of 150 tonnes per cu m is less than half the orbit radius of Mercury—35 million km.

This kind of small-radius star of high-density matter can be created, according to current physics, independently of the supernova process in stellar evolution as shown in Figure 68, which shows an artist's impression of the two currently-accepted paths of star evolution, one of which leads to black shell formation. There is currently insufficient mass in our Sun for it to ultimately evolve into a black shell star, but a star of only a few solar masses bigger does have that potential. It is not a huge difference, so black holes should not be very rare within the universe.

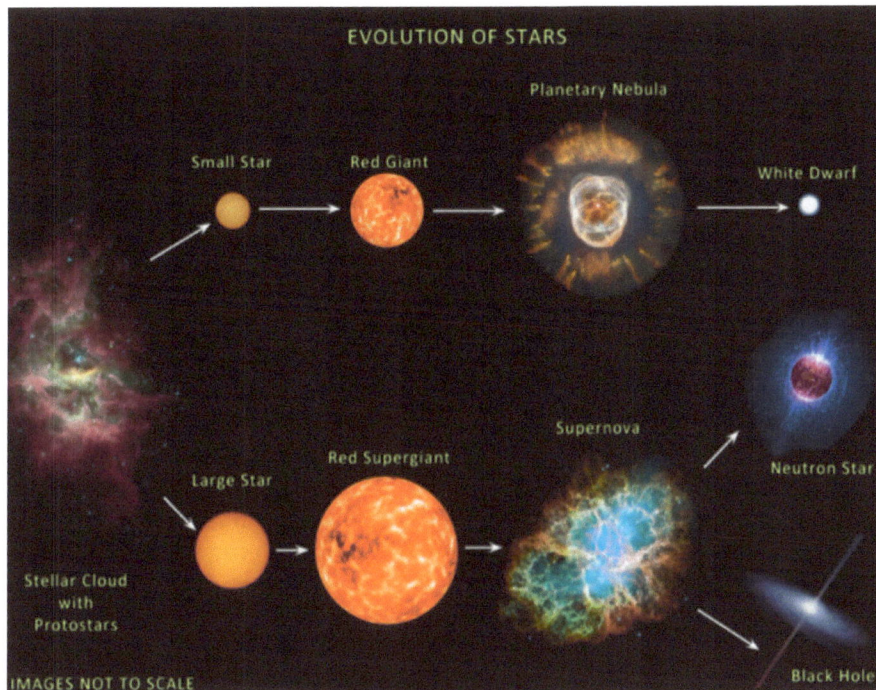

Fig 68. Pre-DOPA view of large-star evolution leading towards the formation of black shell stars (black holes). Courtesy Google.

The conclusion is that it takes only realistic proposals to obtain the escape velocity conditions for the formation of a black shell star without having to invoke relativity and the collapse of that core into a singularity, which is not what the author considers to be practical physics. (It is mentally tiring to hear seemingly-endless physics proposals supported by the statement that, of course, the normal laws of physics will not apply to that particular far-fetched proposal.) The strength of absorption theory, as a refreshing change, is that it is constructed around standard physical concepts and rules.

There is also the confusing aspect related to the talk of the escape velocity of light. Light has to be massless in order to accelerate instantly to its limiting upper velocity of C. Physicists also choose, when it suits them, to argue that, perhaps light is not a wave phenomenon, but a particulate one. These also have to be massless particles. And, if massless, how is gravity supposed to affect them? Moreover, what, exactly is 'escape velocity'? Many people accept it at face value whilst others, who investigate it, become confused. The author has inserted the following description of 'escape velocity' from Wikipedia, which should be entitled, 'ballistic escape velocity'.

Quoting Wikipedia: https://en.wikipedia.org/wiki/Gravitational_potential#Potential_energy

*"In physics, **escape velocity** is the minimum speed needed for an object to escape from the gravitational influence of a massive body.*

The escape velocity from Earth is about 11.18 km/s (6.951 mi/s; 40,270 km/h; 25,020 mph) at the surface. An object which has achieved escape velocity is neither on the surface, nor in a closed orbit (of any radius). With escape velocity in a direction pointing away from the ground of a massive body, the object will move away from the body, slowing forever and approaching, but never reaching, zero speed. Once escape velocity is achieved, no further impulse need be applied for it to continue in its escape. In other words, if given escape velocity, the object will move away from the other body, continually slowing, and will asymptotically approach zero speed as the object's distance approaches infinity, never to come back. Speeds higher than escape velocity have a positive speed at infinity. Note that the minimum escape velocity assumes that there is no friction (e.g., atmospheric drag), which would increase the required instantaneous velocity to escape the gravitational influence, and that there will be no future sources of additional velocity (e.g., thrust), which would reduce the required instantaneous velocity.

For a spherically symmetric, massive body such as a star, or planet, the escape velocity for that body, at a given distance, is calculated by the formula

$$v = \sqrt{\frac{2GM}{r}}$$

Where v is the velocity [in m/sec] G is the universal gravitational constant (G ≈ 6.67×10⁻¹¹ m³·kg⁻¹·s⁻²), M the mass of the body to be escaped from [in kg], and r the distance from the centre of mass of the body to the object [in m].-The relationship is independent of the mass of the object escaping the massive body. Conversely, a body that falls under the force of gravitational attraction of mass M, from infinity, starting with zero velocity, will strike the massive object with a velocity equal to its escape velocity given by the same formula."

What the quotation fails to say is that to travel away from a planet or star, one does not need to reach the ballistic escape velocity. As long as an object, such as a spaceship, has sufficient fuel to keep raising itself off the surface, it can proceed at any slow speed that it chooses. Also, the strength of the gravitational field decreases with distance from a cosmic object, and so the spaceship's fuel consumption will actually decrease away from the body. People often fail to appreciate this. Of course, this raises the discussion point of exactly what powers light in its travels across the universe, and, in particular, in its travel away from a star's surface. Does light really need to travel at its maximum speed to escape a black shell star (black hole)? Or, might it be that within an environment where it is about to be turned back, the speed of light might be locally reduced just as it is when it travels through glass or water? And, further, in saying that a black hole has a black shell where the gravitational pull stops light escaping, we are proposing that the force of gravitation directly controls massless light. This is all quite confusing.

Important note.
The reader must bear in mind that the consideration of black shell stars is being included in this book for purely academic interest. DOPA theory, being a theory for how gravitational force is created, is not concerned with how black-shell stars are formed, or how gravity affects or does not affect light. It is, however, an important part of testing that a theory can be fitted into associated concepts such as black shell stars.

This section assumes that, somehow, gravity does affect light and can exert a drag force on it.

General considerations make that proposal sound nonsensical, but for the purposes of this discussion, the author is taking that unlikely proposition as being true. In this theory, it is proposed that, wherever, in far space, the gravitational wave flux is pristine and omnidirectionally balanced and equal, then gravitational waves have no net effect on light. However, where they are not balanced, within or relatively near to masses of matter, then they do have a gravitational effect on light.

Earlier in this book, some experimental data published in 2015 by Rancourt and Tattersall are mentioned which appear to prove that light can affect gravity, and therefore, gravity may affect light. This experimental laboratory work does, impartially, support the Roberts theory since differential opposing partial absorption theory is the only one that can explain Rancourt's experimental results. Until the development of DOPA theory, Professor Rancourt recorded his results but was unable to explain how his apparently-contradictory results could have been created. The partial-absorption theory of gravitational waves has provided a satisfactory explanation.

The creation of light within a star

Before proceeding further, the author will subsequently refer just to 'light' in order to save repetition, but any comments about light also refer equally to all EMF radiation such as X-rays, microwaves, radio waves, gamma rays, and so on. They are all Electro Magnetic Field radiations, and all travel at 'the speed of light' and all behave essentially in the same way. So, from this point, the reader should note that 'light' is synonymous with 'EMF radiation'.

Consider the inner zones of a typical star. Photons of light are produced within that core zone and fight their way up to the surface. Why? Why do they fight their way up to the surface? And why does it take up to 100,000 years for them to do so? This requires the reader's consideration. Nonetheless, it is generally believed that, at thousands of degrees Celsius and under huge pressure, light can hardly fight its way out. But, over tens of thousands of years, it does. This is somewhat analogous to a pinball table. The ball knows which way it should be heading but is repeatedly intercepted by springs, paddles, and other obstacles that keep diverting it and returning it back up the sloping board.

At the time of printing, there is interesting additional detail at the following web page, but such web pages can 'disappear' as time passes; the author hopes that this will not be one of them. If it is unavailable, then the reader can conduct a simple Google search to find out more.

http://www.abc.net.au/science/articles/2012/04/24/3483573.htm

When stars are normal stars, this migration process eventually results in the light wriggling its way to the star's surface and then being emitted from the photosphere to travel outward as part of the 'solar wind' radiating out from the star. This is what happens with our own star—the Sun.

Of course, some purist theoreticians may argue that the photon that arrives at the star's surface is not the same photon that was created in its core millennia previously; it simply transmutes repeatedly. Nonetheless, this makes no difference to the fact that photons travel outward towards the surface. That is the point relevant to DOPA theory.

Does the light move as a statistical 'Brownian' phenomenon, or is it trying to move outward as a cork analogously rises in water, in response to the density structure of the matter in the star, or its gravitational field? Alternatively, does the excess amount of trapped light form its own 'pressure' structure? Or, does light, naturally, become lost at the outer surface of the star, and thus become replaced by light that is near to the surface, thus forming the equivalent of a chemical concentration gradient which commonly causes chemicals to migrate within aqueous solutions, for example?

At what velocity is light travelling within the star? Generally, it travels more slowly than 'C' within a medium, so, inside a very super-dense medium such as a star's core, how much might it slow down? And even then, how could it realistically take hundreds of thousands of years to travel the radius of a star?

Another question is, "Is light that is created really resorbed by the matter within a star, such that most of it never actually reaches the surface? After all, light that enters our oceans becomes absorbed entirely by the time it has travelled one thousand metres or so downward. If not, what actually happens to it as more and more light is created within a star?

We would all like answers to these questions, but we are obliged to concentrate on how the light behaves within the assumed structure of a star and, in this case, particularly a black shell star.

To conclude this discussion, Figure 69 shows how the author proposes the internal structure of a well-developed 'Light Phase 3' black shell star (black hole) would look under DOPA conditions.

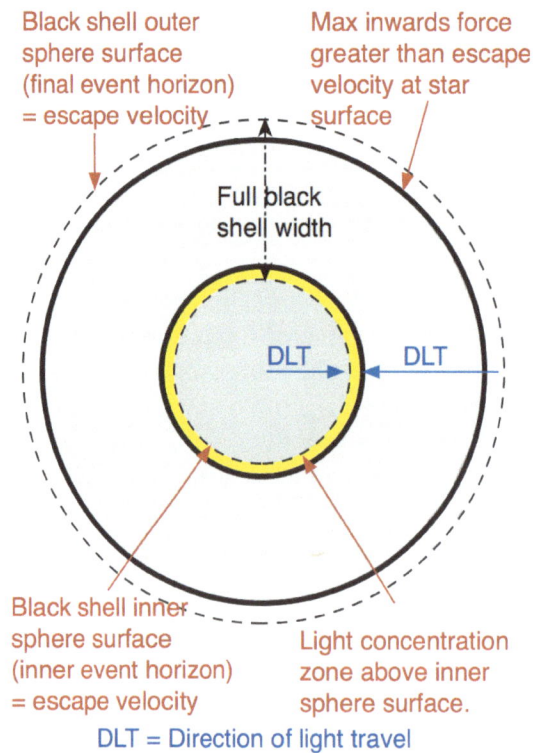

Fig 69. Cross section of a Light Phase 3 black shell star showing the star's physical core and outer surface in bold lines. Limits of primary black shell shown in dashed lines, and light concentration shell shown in yellow.

In Figure 69 one can see that there is a definite, logical, and most-interesting, proposed structure to a 'Light Phase 3' black shell star. Looking at Figure 67, one can see that the black shell is relatively wide, extending from its inner face within the core to the outer face just off the surface of the star. This previously-unsuspected width of the black shell is highlighted in Figure 69 where the physical surface of the core and the surface of the star are drawn in heavy, bold, black lines. The inner and outer surfaces of the black shell are denoted as dashed lines. These match up with those shown in the green graph of Figure 67.

In order to better understand what is happening within the star, we need to look at it one section at a time. We start with the very dense core of the Light Phase 3 star illustrated in Figure 69.

Internal core zone.
This core is extremely dense. Light is generated there as well as elsewhere within the star and tries to work its way out of the centre towards the star's surface.

When light is created in the core and migrates upward, it eventually meets the inner surface of the black shell. At this point, the inner face of the black shell has an inward gravitational force equivalent to the ballistic escape velocity of light, and therefore, any light created within the core and trying to escape cannot pass this barrier. It is the gravitational equivalent of an atmospheric inversion point. When light encounters the barrier of escape velocity, it is turned back. Consequently, the internally-created light accumulates at this spherical horizon and simply builds up here. What, exactly, the consequences of this might be in practical terms are difficult to imagine. Further reflection is undoubtedly called for.

Internal black shell zone.
Within the black shell zone, any light created is drawn down to the light concentration zone above the inner face of the black shell. Also, at the top of this zone, at the surface of the star, the force of gravity reaches its peak value. This is annotated on Figure 69 as the "Max inwards force'.

External black shell zone.
Figure 69 shows that, when the outer sphere moves off the surface of the star, there is the formation of an external shell within which any light travelling tangentially to the star is captured and not released. There is no light emanating directly outward from the star because all light from the star itself is held in by the wide internal shell.

This is very different from the Schwarzschild concept that, somehow, light is emitted from a black hole surface (whatever that might be) and rises up off the surface of the black hole to somehow bounce off an 'event horizon' that exists at some considerable distance away from the black hole. If there is a singularity in there, then why would there be any radiation for the 'event horizon' to keep in? No one currently conjectures as to what happens to all that 'stopped' light or where it comes from. If there is insufficient gravitational force to prevent it reaching the event horizon, then there is insufficient force to drag it back down to the star's surface (singularity?), so does it just accumulate immediately under the event horizon? Of course, theoretical physicists will laugh at this because their construct of the Schwarzschild singularity and its associated event horizon is mathematically theoretical and does not require practical answers. The normal laws of physics do not seem to apply within a Schwarzschild black hole.

Figure 70 shows the much-more-realistic cross-section of a 'Light Phase 3' black shell star illustrating the off-star light capture zone existing away from the star and its black shell surfaces.

In that figure, we can see that there is a new concept—that of an outer black shell. This is significantly different from the current view of an event horizon, in that the outer black shell does not involve controlling light emitted by the star, but instead controls any tangential light that would otherwise pass by the star. This outer black shell is situated at that particular distance where the 'lensing' effect on passing light can just bend it enough to force it into a spiralling orbit and, ultimately, be pulled into the star's surface and into the star.

Thus, despite the fact that this outer black shell does not control the star's own light, it nonetheless creates the boundary that will be observed as being black.

Outside the outer black shell, any tangential light is lensed but allowed to pass by. Also, beyond the outer black shell, there is the reducing gravitational force that can still capture fast or slow-moving matter of any kind in the star's vicinity. (The dissipation of gravitational force within that zone reduces according to standard DOPA gravitational field principles as described earlier in this book.) By means of this mechanism, the black shell star continues to acquire more matter and to grow. If the black shell star has developed as part of a binary star pair, then the black shell can feed off the other star by developing a constant stream of matter and energy from one to the other in a Roche-style destruction procedure.

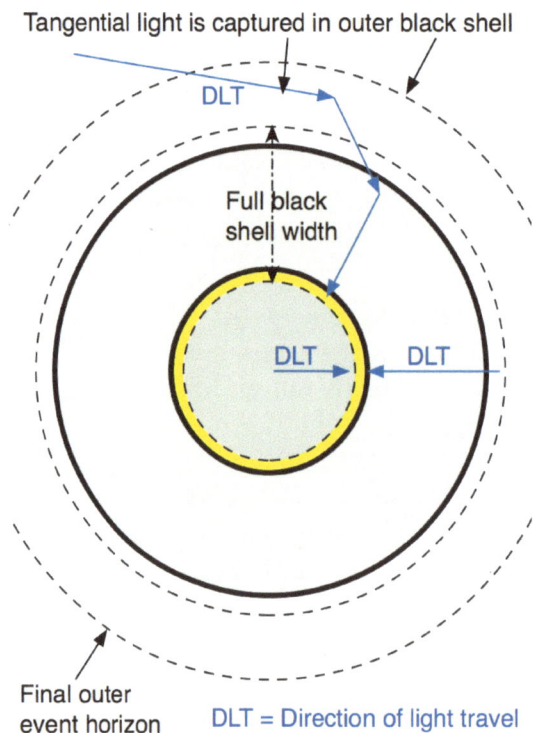

Fig 70. Cross section of a black shell star (black hole) showing the outer black shell in which tangential light is captured and spirals down into the star and is then dragged down to the light concentration zone. Light Phase 3.

92

None of the above descriptions of a black shell star's structure involves the invocation of relativity, spacetime, or singularities. The author's black shell star proposals are constructed on straightforward physics principles.

And yet, it has to be pointed out that the whole picture of whether or not black shell stars (black holes) actually exist, is (taking DOPA theory) dependent on gravity being able to affect light. Professor Rancourt's work, described in more detail below, provides evidence that it does, and that there is no need to use the concept of spacetime to justify gravitational force affecting light. The interesting point about Rancourt's work is that he was unable to explain the results of his experimental work at all since it seemed contradictory. However, when the author applied his differential absorption theory to the results, they were explicable. That is a significant supporting factor for DOPA theory. Also, the proposals in Section 5.5 of this book suggest that nature may ultimately prevent the theoretical development of a singularity from happening. That would be fortunate.

What happens to gravitational waves when they enter a black shell star?

The final question that has to be put concerning gravitational waves and black shell stars is what happens to gravitational waves when they enter a black shell star? The answer, according to absorption theory, is that, in principle, gravitational waves can enter and leave a black shell star through its shells without any effect. It is only in the most extreme cases, where the mass of the star is sufficient to totally absorb gravitational waves, that they enter but do not go out.

The reason for this is that, outside matter, gravitational waves are unaffected by themselves.

Even within matter gravitational waves are unaffected by each other. Just in the same way that light in space does not interfere with other light, gravitational waves do not interfere with other gravitational waves. So, both entering and leaving the various shells above the surface of the star, it is expected that gravitational waves travel at the speed of Cg and are not accelerated on entry or retarded on exit from the black shells around the star. This conforms to Roberts' proposed third postulate to the Special Theory of Relativity. Once the waves enter the star itself, however, their velocity may slow to less than Cg.

Nonetheless, irrespective of the fact that high gravitational forces are being exerted on the atoms of the constituent matter of the black shell star, gravitational waves within the star's mass are not being affected by other waves. There is no fundamental difference in terms of their behaviour. Some new and very interesting propositions have arisen, however, which are described and discussed in sections 5.4 and 5.5 below.

The above reasoning explains why the author eschews Einstein's 'shoelaces' circular reasoning. Einstein says that gravity warps spacetime and, having created the warp, other matter is obliged to follow the warp gradient. This is like trying to lift oneself up by pulling on one's own shoelaces. Einstein says that matter must abide by the rules of its own creation. The author, on the other hand, says that gravitational waves create gravitational force and so cannot be affected by their own creation. To believe otherwise would be to believe in a physical paradox. Einstein can afford to believe in his paradox because his explanations are entirely theoretical.

4.13 Can this theory explain why orbiting objects such as satellites do not slow down as they continually run into oncoming gravitational waves that could create drag?

Yes, it is quite straightforward to explain why this does not occur, this explanation being unique to the author's theory for the creation of gravitational force. In a nutshell, because of his proposed third postulate to the Special Theory of Relativity.

This is one of the most significant of the questions presented in Section 4 because it is the reason why the centuries-old, parallel concept of aether particles moving through outer space, striking cosmic objects, and creating a consequent force on them, was rejected.

The author's concept and explanation successfully reject the objections raised in the old particulate concept. Of course, absorption theory does more; the use of gravitational waves that are differentially absorbed allows the theory to relate the force of gravity developed at the surface of any cosmic object to its mass and radius directly. This is fundamentally important.

The summarily-rejected 'particle concept' was highlighted in Lecture 2 of the well-known November 1964 series of lectures given by Professor Feynman at Cornell University in the USA.

http://www.cornell.edu/video/richard-feynman-messenger-lecture-2-relation-mathematics-physics

It was thought, at the time of the 1964 lectures, that it would be impossible for planets and even man-made satellites to remain in stable orbits in the long term if they were to be slowed down by striking extra particles (as if running into the rain as Feynman put it), and this was a sensible point of view. And that is why Fenmen rejected it.

However, DOPA theory has removed this obstacle in this piece of original work which has derived additional strength from the separation of the old from the new theory.

In 1978, the author developed his own new concept that gravitational waves might perform the same function but found that his theory at the time was not fully developed. It depended on the proposed existence of gravitational waves (which was not a unique proposal) but could not progress further because he had not, at that time, developed the concept of opposing differential absorption. That was the key to the present theory and preparation of this book. Consequently, the author decided to wait until the existence of gravitational waves could be verified. After the confirmation by the LIGO Caltech/MIT experiments, which confirmed the existence of gravitational waves, the author decided to continue the development of his paper, and this work is the outcome.

Let us first examine what happens with light, since we are familiar with it. Although an object may move through space, the light landing on its surface does so at the speed of light, C whether it arrives at the front or back. The same applies to gravitational waves which propagate at the speed of light with a velocity Cg. If we examine Figure 71, we can see that, despite the Solar-reference-frame orbital velocity of V km/hr, we know that light still arrives at its facing and lee surfaces at exactly C, the speed of light in vacuum. Since gravitational waves, although fundamentally different from EMF waves, travel at the speed of light in the same way, then they will also arrive at the surface of the subject planet at the same velocity of Cg on both the facing and lee surfaces.

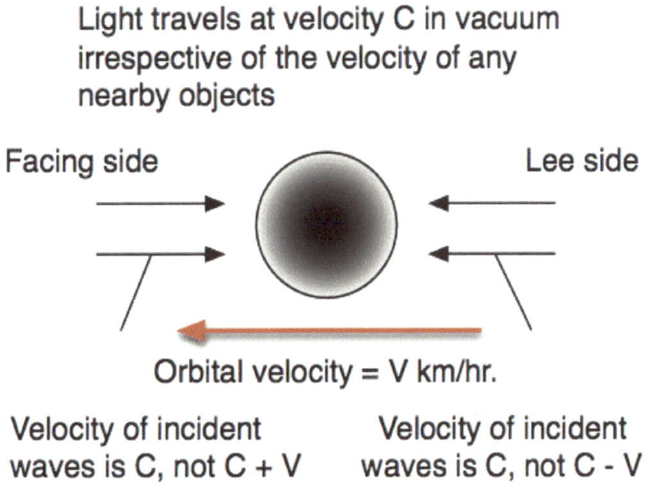

Fig 71. Light and gravitational waves arrive at the leading and trailing surfaces of a moving object at the same speeds C and Cg.

We can, therefore, understand that, since the velocity is the same on both sides, there is no additional drag created in either direction to unbalance the differential absorption of oncoming and rear-coming waves.

The apparent difference in velocity is just that, an apparent difference. When viewed from the planet's frame of reference, it is stationary, and this is the reason for the waves' identical incident velocity on its surfaces in all directions.

As an incidental outcome to this work, the author proposes that there should be added a third postulate to the special theory of relativity:

3. The speed of gravitational waves (Cg) in a vacuum is the same for all observers, regardless of the relative motion of the observer.

4.14   Does this theory have outcomes that match and support Einstein's theories of relativity?

Yes, DOPA theory has no disagreements with relativity apart from its use of spacetime to substitute for gravitational force. This book proposes that special relativity can use the three-dimensional concept of this new gravitational theory to substitute for the mechanism of warped spacetime. Prior to this point in time, there was no other alternative theory upon which relativity could rely. Now there is.

Relativity is all about how matter behaves in the universe, which is attributed to its response to spacetime. This differential opposing partial absorption theory does not address responses as does relativity because DOPA is a theory that only explains how gravitational force is created. Behaviour in relation to forces is only discussed in this book to show that absorption theory does not contradict what we observe and therefore does not contradict relativity.

4.15   Does this theory contradict the work of Newton or Einstein?

Insomuch as neither of them was able to propose a rational working mechanism for the creation of gravitational force, it does not. Insomuch as it says that the lack of an explained gravitational mechanism was the downfall of Newton's work, and is the weakness in Einstein's work, it does.

DOPA theory proposes that Einstein's adopted concept of spacetime being a four-dimensional state of reality is unacceptable because there is no scientific justification for it. It may be an explanation for how matter behaves, but that does not mean we should confuse it with our reality. It also contains an inexplicable circular contradiction that the very existence of matter warps spacetime and then, when matter has warped spacetime, it is obliged to travel along spacetime in response to its own distortion. This is akin to trying to pull oneself up by one's shoelaces. Circularly-contradictory and, hence, implausible.

Einstein co-opted spacetime from Minkowski's paper of 1908—who, himself, had copied the idea from Poincaré in 1905—in an attempt to explain gravitational force, but for all the thought he could give it, he was only able to come to the conclusion that gravitational force did not exist, and so he used spacetime on which to found his theories. And yet, puzzlingly, he admits that there is a gravitational field and he writes of gravity waves and gravitational waves in his theories.

Einstein laid out his excellent forecasting mathematical work on the basis of theoretical spacetime, but it is reasonable to propose that he could have done the same on the basis of DOPA theory, had it been in existence at the time.

Absorption theory provides a rational explanation for how gravitational force is created without having to invoke a fundamental change in the reality of our three-dimensional universe. It also provides a rational explanation for the existence of gravity zones, and improves on this because Roberts' theory can be extended down into the planetary or stellar mass creating the gravitational force without contradicting itself with matters such as time dilation within the object's varying gravitational field (in Einstein's case), outward-acting gravitational force at the centre of the object (in Newton's case). It also addresses the problem from which Newton and Einstein both suffer—explaining how their theoretical structures transmit information within their fields and how their fields react at huge distances with their electromagnetic information transmission exceeding or even acting at the speed of light.

As a contribution to relativity, absorption theory provides a working mechanism for the creation of the long-sought-for universal gravitational field and potential gravity zones. It is constructed on the basis of consistently-inwards travelling gravitational force within matter. It also explains how gravitational force is potentially present around objects, so that gravity zones are only tapped into when matter is present within them.

In the same way as relativity, absorption theory fits the many conditions of cosmic behaviour against which it has been tested so far.

4.16   Does the absorption theory mean that gravitational waves must be getting weaker throughout the universe?

This is unlikely, for two reasons:

Firstly, we already know from the LIGO work that there are new sources of gravitational waves. These could make up for any absorption loss.

Secondly, and relevant to that point above, the amount of matter in the universe makes any loss infinitesimally small. Replenishment is, therefore, more than likely to replace any losses.

In addition to matter's intrinsic transparency to gravitational waves, there is so little matter present, volumetrically, in the universe that it forms virtually no obstruction to the ubiquitous gravitational waves. Thus, there is no reason to consider matter to be a significant depleting agent in diluting the overall cosmic flux density of the gravitational field.

Consider that the volume of the main planets within our Solar System-are as follows:

Volumes cu km
Sun 1.49E+18
Mercury 6.09E+10
Venus 9.29E+11
Earth 1.08E+12
Mars 1.63E+11
Jupiter 1.43E+15
Saturn 8.27E+14
Uranus 6.83E+13
Neptune 6.25E+13

Total volume of matter is 1.49E+18 cu km + 10% for unknowns = 1.64E+18 cu km.

Take the radius of Neptune's orbit as an arbitrary radius to define the volume of space occupied by our planets.

The radius of Neptune's orbit is 4.5 billion kilometres (2.8 billion miles)
The volume of Neptune's orbital sphere is 3.82E+29 cu km.

The ratio of matter to the solar system's effective volume is, thus, very approximately 1 to 1 trillion.

And, of course, this minuscule ratio is for the highly-populated Solar System, completely ignoring the empty space between stars and the even more immensely empty inter-galactic spaces beyond.

There is, clearly, insufficient matter to deplete, by absorption, the overall gravitational energy available within the universe by any significant amount over any relevant time scale. That leaves virtually an infinite amount for humans to tap into as an energy source.

4.17 Does the DOPA absorption theory offer an explanation for the unexpected and previously-unexplained results of a tree growth experiment conducted on the International Space Station in 2009?

Yes. A novel tree-growing experiment was conducted onboard the International Space Station (ISS), sponsored by the Canadian Space Agency assisted by NASA during 2009. Near the end of his scientific career, dedicated to understanding how trees produce wood, this experiment was proposed and designed by Professor Rodney Savidge of the University of New Brunswick, Canada, Principal Investigator of the APEX-Cambium research project.

To appreciate the purpose of the Apex-Cambium experiment one needs to be aware of some of the known features of the growth habits of tree trunks and branches.

Firstly, during their primary or elongation growth, flowers and new shoots tend to grow towards the light. The heads of some flowers and elongating stems actually track the Sun during its daytime sky transit. Woody perennials such as shrubs and trees are not different except that their hardened form at the end of each growing season displays the net cumulative annual growth response. Overall, the central axes of trees grow vertically, irrespective of the Sun's position. For example, towards the arctic circle, where the Sun never reaches higher than forty-five degrees above the horizon all year-round, the trunks of trees are still perfectly vertical (R. Savidge, personal communication). If one thinks about it, we accept that as the normal thing for trees to do. The vertical growth of trees everywhere on Earth is well-recorded and is called 'negative geotropism'. The term 'negative' means that the growth defies gravity and thus proceeds upwards.

This strong tendency for a tree to achieve a vertical axis is not just a matter of the leading tip of the tree growing upwards, but occasionally is misconstrued as merely reaching for light. On over-steep slopes where the soil and tree root systems are subject to slipping, the trunk at the base, near the roots, actually bends into a curve, thus keeping the tree trunk above vertical in conformance with the verticality of gravity. The bottom metre or so of a tree trunk on a steep slope is not uncommonly found to be at a curving angle, whereas the trunk above is vertical! This well-documented phenomenon of trunk righting is not achieved solely by primary growth, but also requires secondary or diameter growth in the cambium.

As explained by Professor Savidge, by means of both primary and secondary growth, branches grow out sideways at angles particular to each kind of tree, rather than vertically, in order to increase the amount of light capture in support of photosynthesis. A branch is a cantilevered structure and, thus, it seems natural that they would grow special features to support themselves, not only their weight but often also the additional weight of ice and snow and against the force of wind.

Secondly, as well-informed laypersons know, there are different types of wood present within the trunks and branches of trees. Most of us think, simplistically, that tree trunks and branches comprise wood and bark and that where branches meet the trunk, the wood grows extra-strong 'knotty' wood to support the long-lever stress caused by the weight of those branches. That seems to be logical, but it is not such a seemingly-simple case.

What actually happens is that, not only are branches supported by knotty wood where they meet the trunk, but wherever a component of a tree is not upwardly vertical, a special 'reaction wood' forms to reinforce the structure. In most broad-leaved species, a tension-resistant strong wood consisting of gelatinous fibres and called 'tension wood' forms along the upper length of branches, as shown in Figure 72.

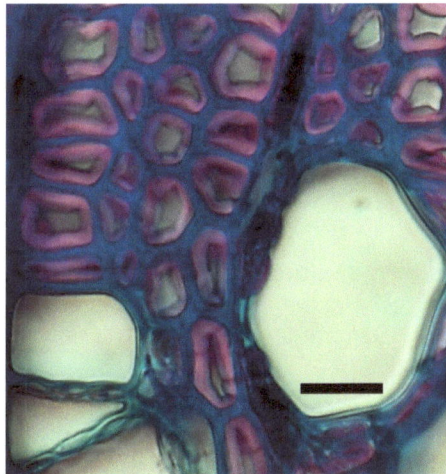

Fig. 72. A microscopic cross section of tension wood showing the pink-stained gelatinous (cellulosic) layer within and diagnostic of the gelatinous fibre. The bar is 20 μm. (Photo courtesy of R. Savidge)

About a century ago tree scientists thought that, if they bent living willow stems into vertical loops, then all around the outside of each loop, the tensile forces would cause tension wood to develop. It was to be expected if the mechanism was activated by imposed stress. But, the experimental results did not show this. Amazingly, the change from normal wood into tension wood formation occurred only at the upper and lower sector of the loop and there, only at their zenith positions, as illustrated in Figure 73.

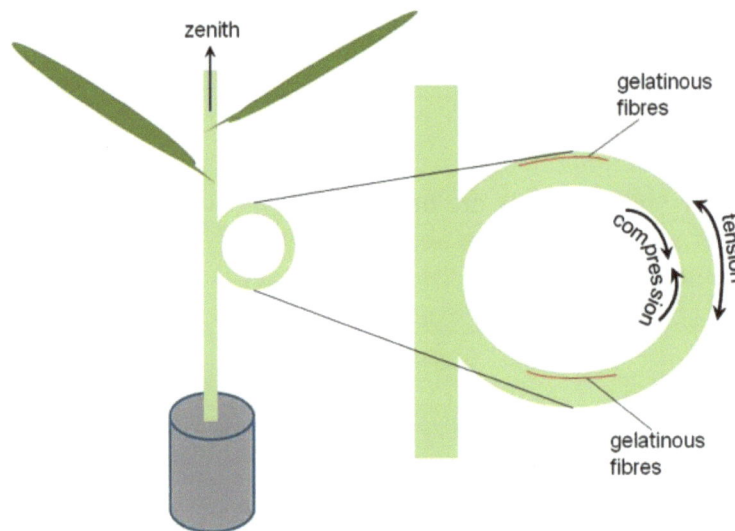

Fig. 73. Schematic of a willow stem showing the localised and unpredicted bottom-most position where tension wood gelatinous fibres are produced in a compression zone after the stem is looped and grown for several weeks as a looped stem. Also, showing the lateral positions where tension fibres are, surprisingly, not produced at all. (Courtesy of R. Savidge)

What happened has to be described carefully. At the topmost part of the stem loop, tension wood was formed along the upper surface while normal wood was produced along the lower surface. This was in accordance with what was known to occur along branches. After all, the top of the outermost part of the loop is under tensional stress whereas the inside of the loop is in compression. But what was very surprising was that, at the lower or bottom part of the loop, tension wood was formed along its upper or zenith surface where cambium and wood were in compression, and normal wood was formed along the lower surface where the wood was in tension. And, at the curving but more-or-less vertical sides of the loop, where tension and compression stresses were still continuously present, surprisingly, no new type of wood grew.

Savidge recognised that this unusual development could be because cambium cells respond to something about gravity, as distinct from the force of weight that is usually equated to gravity, not to the strain produced by an imposed stress—not to weight per se.

And, supporting that interpretation, Savidge found that, if a dormant horizontally-oriented branch of several centimetres diameter was removed from a willow tree in late winter, when the cambium was dormant, and lengths were cut from the branch and rotated 180 degrees such that what was their bottom side became the top side, cambium in the segments in inverted horizontal orientation produced a new annual ring of wood. And, of course, these short segments have no imposed tensile or compressional stress on them. After a month of growth under controlled conditions, a considerable new annual ring width was present, and what was the nadir side cambium producing normal wood when intact in the tree had, after rotation into the upper or zenith position, transformed itself into tension-wood producing cambium, with the converse occurring on the

100

opposite formerly tension-wood-producing side of the branch. So, a complete reversal of phenotypes was achieved simply by rotating the segment 180°.

Savidge wondered whether he could advance knowledge by setting up a tree-growing experiment for willows in a weightless environment. It could be argued that, if his hypothesis was correct, then looped stems of saplings, deprived of the strain created by weight on Earth, would grow no reaction wood at all, at any point. The obvious location for such an experiment would be the International Space Station (ISS). So, he persuaded the Canadian Space Agency and NASA to support the concept of his proposed experiment. Savidge reasoned that, rather than a state of microgravity on the ISS, everything within the ISS as well as the ISS itself was actually experiencing 0.9 g and therefore that his willow saplings might respond to gravity the same as if they were on Earth. And, as it turned out, they did! He evidently was right, although Professor Savidge (personal communication) has cautiously emphasized that in keeping with the need for rigorous science, the APEX-Cambium grow-out on the ISS should be repeated to confirm the finding.

The APEX-Cambium grow-out experiment conducted in the ISS, shown in Figure 74, was designed by Savidge and conducted by trained astronauts. The trees 'flew' on the ISS in the same earth-upright zenith-nadir orientation that potted plants have on earth. Fortuitously, the experiment benefitted from the fact that during the one-month November-December 2009 grow-out, the ISS was not accelerated outward (i.e., boosted to a higher orbit) and thus its orbital altitude drifted downward steadily throughout the period.

One important question that would still remain to be answered is exactly how trees in the weightless state are able to sense the direction and force of gravity that holds the ISS in its orbit.

Fig. 74. The actual APEX-Cambium trees on the ISS. Each row of six willow trees was a distinct genetic clone and four of the six trees were looped with the other two serving as un-looped controls. The trees were grown on the ISS for one month, then fixed in formaldehyde solution and return to earth. (Photograph by astronaut Robert Thirsk, courtesy of R. Savidge)

101

Before proceeding further, we need to be sure we understand the state of gravitational and centrifugal forces affecting the ISS and its contents, based on the state of knowledge of gravity back in 2009.

The force of gravity downwards into the Earth is at a maximum at the surface of the Earth (approximately 9.81 m/sec^2) and decreases with height away from the Earth's surface. At the altitude of the ISS, the force of gravity downwards is about 90% of its Earth-surface value. Say, 8.83 m/sec^2. In general terms, the force of gravity is little different at the orbit height of the ISS than it is at the Earth's surface. So, in that case, why do the astronauts and the ISS itself appear to float weightless in space? The answer is that the centrifugal force of the ISS's orbit counteracts almost exactly the gravitational force at that orbit height. And so, objects and people are forced outwards with exactly the same force as they are forced inwards, and they feel and behave as if weightless. This state is never in perfect balance, as one might reasonably expect, and so it has come to be known as a state of 'microgravity' in recognition of the minute variations experienced away from absolute zero gravity. Further, Savidge prefers the term 'microweight' to 'microgravity' in recognition of the fact that, in reality, there is considerable gravity affecting the ISS, it is just that it is counteracted by centrifugal force. Therefore, it is reasonable to be more accurate by referring to the balanced effect as 'microweight'. This is important when researching a subject such as the one in hand.

After all, how can one undertake research on a gravity-related subject if one considers—as a result of incorrect nomenclature—that there is no gravity present?

And, in a nutshell, Savidge's hypothesis was: "If tension-wood formation is a reaction to the pull of gravity, as opposed to weight-induced stress, then tension-wood development will occur at upper positions in loops of trees grown in the weightless environment of the ISS".

The tests took three years or so to design, agree, set up, and implement.

And, amazingly, the professor's hypothesis was ratified in that tension wood <u>did</u> develop in looped willow stems at the top and bottom of the loops as previously described above, but in a microweight environment. Also, the same features developed on some straight stems left at an angle of about 45 degrees to the vertical—additional experimental support for the hypothesis.

But this did not answer the critical question of how does a tree detect the force and direction of 'gravity'? This could not be answered by considering the question in the light of Einsteinian curved spacetime theory or Newtonian attraction theory. In neither theory was there a mechanism for the creation of gravitational force that could be compared to the microweight environment of the experiment and which could, therefore, be used in scientific reasoning and logic to provide an answer.

Immediately associated with that question is the further one of how did the tree stems detect their orientation in a microweight environment that was, for all practical intents, a net gravity-free environment?

And, finally, we come to the question asked of the author by Professor Savidge, "Just as the ISS, though weightless while in freefall orbit, somehow detects earth gravity to keep orbiting, apparently so did the cells within the tree. I am wondering how your theory might explain this?"

And here is the answer:

To understand what is happening in the case of any object adjacent to a planet, whether in orbit around that planet or not, we must recognise that, in accordance with DOPA theory, the planet differentially absorbs opposing gravitational waves and, in that way, constructs a zone of potential gravitational force around itself. Thus, in orbit, the ISS and any persons or experiments within it are subject to the net inward vector of gravitational force created by being exposed to both the pristine incoming waves and the depleted outgoing waves from the Earth. This book has taken pains to describe that process thoroughly. It is the Earth's primary gravity that is neutralised by the deliberate placing of the ISS at such a height and such an angular velocity as to perfectly cancel out the inward net gravitational force. And so, a weightless, net gravitational force environment is created through which the ISS orbits.

But, and this aspect was alluded to in Section 4.9, within any object on a planet's surface or in space, that small, even tiny, object, molecule, or atom is subjected to exactly the same differential absorption process as is the planet. Each object or sapling tree in the ISS is subject to the creation of its own, independent, differential opposing partial absorption experience that creates its own spherical gravitational zone around it. DOPA theory recognises that every object in the universe creates its own proportionate gravitational zone around itself. It is the interaction of one gravitational zone with the other that controls the gravitational behaviour of objects, big or small. This is similar, but on a different scale, to the two spheres in Figure 35 where Body 1 might represent the Earth and Body 2 the ISS. That seems straightforward when considering the Earth and the Moon or similar objects, but is easy to forget when one of the objects is the size of a mere tree sapling. Yes, every object, no matter how tiny is subject to DOPA effects.

Thus, impinging on every tree sapling, there are the incoming pristine waves and the outgoing Earth-degraded waves, but—and this is the most important point in this experimental context— within the tree sapling, those differentials are microscopically added to by the sapling's own wave absorptions and this additional gravitational field creation is NOT cancelled out by the main inwards/outwards force that keeps the ISS in orbit. And, this additional field generation is the same one that would be created on Earth by the same size sapling, because it is the mass of any body that creates its own, individual, gravitational response. It is the same response that is described in Figure 35 and Figure 47, and does not matter whether on Earth or in space, whether in a strong gravity field or an apparently-net-zero gravity field created by centrifugal force counteracting gravity.

The outcome of this train of thought and its conclusions is that any plant growing on the ISS will be receiving and experiencing the same internal 'gravitational signals' as if it were growing on Earth. And this leads to the further conclusion that what all trees are responding to as they grow on Earth is the strength and orientation of the individual DOPA gravitational effects that they are experiencing within their chemical reactions and fluid processes. These DOPA effects will, necessarily, vary from place to place within a tree because the effective mass of the trunk at any point will be significantly different from the effective mass of a branch, or twig, or leaf. If it were

103

just the coarse differential of the Earth's overall gravitational differential field, then subtle differences from one part of the tree to another could not be implemented. By responding to its own internal gravity differentials, the tree behaves exactly as the theory would demand—small internal gravitational forces in a leaf and greater ones in a branch.

In essence, therefore, the tree grows according to its own individual responses to gravitational waves in a way that varies from place to place within its own body. These responses are registered in the internal chemical processes and guide the growth responses of the tree.

DOPA theory, again, meets and explains a question that has heretofore not been capable of solution by consideration of Newtonian or Einsteinian theories. DOPA absorption theory works.

Note: Professor Savidge has suggested to the author the interesting possibility that, because DOPA theory requires 'information' that is abundant throughout the universe and passing through matter essentially unhindered, neutrinos could be the force inherent in gravitational waves. It is a conceptual leap into quantum physics, perhaps deserving consideration but needing research.

4.18 Has DOPA theory produced any new concepts and explanations that have not been proposed by others?

Yes, differential opposing partial absorption theory...

1) has propounded, for the first time, a mechanism for the creation of gravitational force.

2) has extended that mechanism to show how the universal gravitational field is created and

3) has shown of what the universal gravitational field is composed.

4) has revealed how information concerning its status can be transmitted so as to control bodies at significant distances apart. For example, it addresses and satisfies the long-standing problem of what would happen to the Earth's orbit instantaneously if the Sun were to suddenly disappear and before the end of the eight-minute period for EMF information to reach the Earth. No other theory can do this, making it a major step in cosmic understanding.

5) has given rise to a new explanation for Roche theory and

6) has provided a new explanation for shell theory.

7) has posed the remote possibility that magnetic materials may be a source of information concerning materials controlling the behaviour of gravitational waves and thus of gravity.

8) has shown how black shell stars (black holes) may be formed without recourse to theoretical mathematical equations based on relativity, and

9) has described the likely structure of black shell stars logically deduced from the four tenets on which DOPA theory is based.

10) has raised the possibility that the centre of the most massive stars may have liquid or solid cores without inertia or momentum.

11) has provided an explanation for how light can affect gravitational force.

12) It has explained the creation and existence of inertia and momentum.

13) has explained why singularities are not needed to form black shell stars and

14) has proposed a mechanism that would prevent singularities from forming in the core of massive stars.

15) has provided a new concept that there is likely to be a limit to the strength of the gravitational force that nature can achieve.

16) has highlighted the interesting sequitur that there might be a limit to the amount of pressure that nature can achieve.

17) has led to a closer critique of Newton's equation and the development of a modified version of it that allows new examinations to be undertaken.

These seventeen new contributions are all described in Chapter 5.

CHAPTER 5 - HAVE THERE BEEN ANY NEW OUTCOMES FROM THE DEVELOPMENT OF DOPA THEORY?

5.1 The use of the principles of this theory to explain the creation and mechanism of both inertia and momentum, whether linear or rotational.

It is common-sense that if we, as humans, approach a 1 tonne weight (such as a car) suspended on a crane hook, and we wish to move that weight sideways, we expect to perceive a significant resistance to our attempt to move it. Why? And if that car was a new vehicle parked on a horizontal road with the handbrake off, we would expect to experience the same high resistance to our attempt to start moving it sideways. Again, why? If we are moving an object sideways, then the Earth's downward force of gravity is not resisting us or assisting us. And yet we experience a significant resistance. In free fall in space, when objects such as the components of the International Space Station are being assembled by astronauts in space suits, those objects, whilst having no apparent weight, being in free fall, still have their same level of inertial resistance as if they were suspended from a crane on the Earth's surface. Why? And, How?

Absorption theory can now explain, in simple terms, without any additional assumptions, why and how the force of inertia is created and why it is directly proportional to the mass being accelerated and the acceleration to which the mass is subjected.

There are two simple questions that no one has been able to answer prior to the development of DOPA theory: Why does inertia exist? What mechanism creates the force of inertia?

For an impartial view, one can look at Wikipedia, where we find the following opening statement:

"Inertia is the resistance of any physical object to any change in its position and state of motion. This includes changes to the object's speed, direction, or state of rest. Inertia is also defined as the tendency of objects to keep moving in a straight line at a constant velocity. The principle of inertia is one of the fundamental principles in classical physics that are still used to describe the motion of objects and how they are affected by the applied forces on them."

https://en.wikipedia.org/wiki/Inertia

There is no mention that it is caused by gravitational force. Moreover, there is no mention of the concept of acceleration. When an object has its position or velocity changed, it must necessarily have been subject to not just a force, but an acceleration. The force must have been exerted over a period of time, no matter how long or how brief. A careful search in Wikipedia finds no explanation for how inertia is created—just how it behaves.

And, while creating the First Law of motion, Newton did not use the term 'inertia'. Instead, he did what he was used to doing and what he had done in his concept of gravity. He described something without giving any explanation for its working mechanism. He said that the phenomenon was caused by natural forces inherent in matter which resisted any acceleration". Typical of these geniuses—describing it should suffice.

Wikipedia also describes what an ultimately unsuccessful struggle Einstein had in trying to come to a unified theory including inertia and momentum. Read the following extract from the above Wikipedia URL reference and note the problems and caveats built in. Not the least of which is the existence of spacetime theory, which DOPA theory challenges.

"Albert Einstein's theory of special relativity, as proposed in his 1905 paper entitled "On the Electrodynamics of Moving Bodies" was built on the understanding of inertia and inertial reference frames developed by Galileo and Newton. While this revolutionary theory did significantly change the meaning of many Newtonian concepts such as mass, energy, and distance, **Einstein's concept of inertia remained unchanged from Newton's original meaning (in fact, the entire theory was based on Newton's definition of inertia).** *However, this resulted in a limitation inherent in special relativity: the principle of relativity could only apply to reference frames that were inertial in nature (meaning when no acceleration was present). In an attempt to address this limitation, Einstein proceeded to develop his general theory of relativity ("The Foundation of the General Theory of Relativity," 1916), which ultimately provided a unified theory for both inertial and non-inertial (accelerated) reference frames.* **However, in order to accomplish this, in general relativity, Einstein found it necessary to redefine several fundamental concepts (such as gravity) in terms of a new concept of "curvature" of spacetime,** *instead of the more traditional system of forces understood by Newton.*
As a result of this redefinition, Einstein also redefined the concept of "inertia" in terms of geodesic deviation instead, with some subtle but significant additional implications. **The result of this is that, according to general relativity, inertia is the gravitational coupling between matter and spacetime.**
When dealing with very large scales, the traditional Newtonian idea of "inertia" does not actually apply and cannot necessarily be relied upon. Luckily, for sufficiently small regions of spacetime, the special theory can be used and inertia still means the same (and works the same) as in the classical model"

The reader can note that to come to some accommodating explanation for inertia, and to explain it, Einstein had to create his new invention 'curved spacetime', ignoring the fact that he had no explanation or mechanism for the existence or working of spacetime and its alleged warping by matter.

The reader can note the 'accommodations' that Einstein had to accept in his struggle to make progress. He had to redefine several fundamental concepts which he would not have had to do if he had had DOPA absorption theory. He had to redefine inertia in terms of geodesic deviation. This produced significant additional implications that he then had to address.

Finally, he came up with the strained definition that inertia is the gravitational coupling of matter with spacetime. So here, again, is an example of Einstein doing his circular reasoning to get out of a hole. He invented spacetime to get out of the first hole and then used spacetime to get out of the second, consequent, hole. This is the same as when he says that 'matter in general' distorts spacetime and then says that 'matter in general' must follow the distortions in spacetime that it, itself, generically makes. It is not difficult to see that, although relativity works, it was a bit of a cobbled-together concoction. And without spacetime, in the first place, had he been using absorption theory, Einstein could have progressed on to develop his theory of relativity much more quickly and more easily.

This is the same struggle that the author experienced, *but in his case, he was able to make the breakthrough* without having to resort to the steps taken by Einstein, each one carrying ongoing problems that, in turn, needed further explanations. Absorption theory, on the contrary, contains an explanation of all the mechanisms necessary, is self-contained and does not carry the need for further explanations. It is complete and self-sufficient. Which is why it is proposed as an improved foundation for general relativity.

Finally, at the same reference URL as above, Wikipedia comes down to the truth of the matter, when it states,

*"Newton's original ideas of "innate resistive force" were ultimately problematic for a variety of reasons, and thus most physicists no longer think in these terms. **As no alternate mechanism has been readily accepted, it is now generally accepted that there may not be one which we can know**, the term "inertia" has come to mean simply the phenomenon itself, rather than any inherent mechanism. Thus, ultimately, "inertia" in modern classical physics has come to be a name for the same phenomenon described by Newton's First Law of Motion, and the two concepts are now considered to be equivalent."*

The reader's attention is directed to the Wikipedia words highlighted in bold immediately above.

"As no alternate mechanism has been readily accepted, and it is now generally accepted that there may not be one which we can know..."

Prior to the development of absorption theory, there has been no acceptable explanation of a mechanism for the creation of the resistive force of inertia. So persistent has been this lack, that people seriously consider that there may not actually be a mechanism. However, a strict logical examination of any such assumption produces the outcome that if there is a phenomenon, there must be a mechanism for its creation. Nothing just 'exists'.

So, what does absorption theory say about inertia?

Figure 75. demonstrates that we are familiar with the net downward drag of gravity to create the phenomenon we think of as 'weight'. In that figure, the two arrows at '1' show the powerful, pristine, incoming gravitational wave from outer space and the outgoing much-reduced gravitational wave which partially neutralises the incoming wave. These two sets of waves are deemed to act within the 1 tonne weight to create gravitational force. The single arrow at '2' represents the net downward vector of gravitational force thus created within the 1 tonne weight.

What we do not think about so much is that the weight is derived from the continuing attempt by gravity to accelerate us and other objects downward towards the centre of the Earth with a desired acceleration of 9.81 m/sec^2. In short, <u>weight is caused by</u> a continuous state of **attempted** but <u>prevented</u>, acceleration. Also, we recognise that once the vertical resistance is removed—if the spring balance breaks, or if we step off the top of a tall building—acceleration now takes hold; we become weightless, and we experience an increasing vertical velocity downward until we meet the floor below and are almost instantaneously decelerated to a standstill—which will damage us severely or kill us.

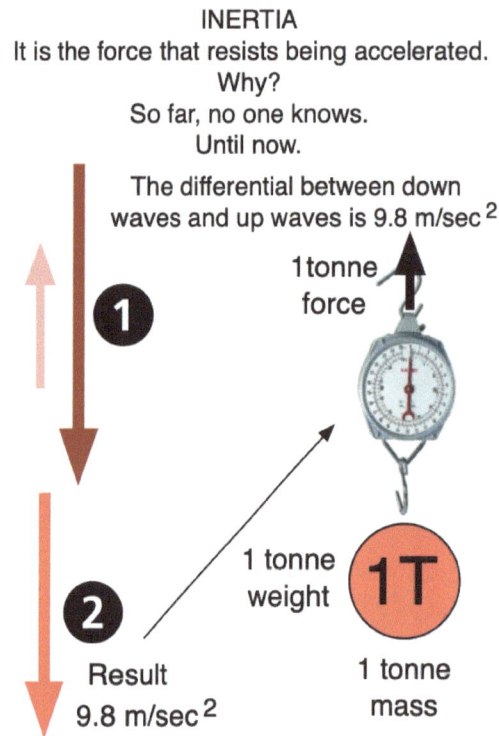

INERTIA
It is the force that resists being accelerated.
Why?
So far, no one knows.
Until now.

The differential between down waves and up waves is 9.8 m/sec^2

1 1 tonne force

1 tonne weight

2 Result 9.8 m/sec^2

1T

1 tonne mass

Fig 75. Differential absorption creates the downward gravitational force to load a 1 tonne weight suspended on a weighing balance.

So, in the case of us and gravity, we stand on the floor and the resistance of the ground acts against the net gravitational force acting vertically through us. And, apparently, once resistance is removed, the force continues to act, but ceases to be felt, and we consider ourselves to be in a state of 'free fall' because every atom in our body is now being accelerated successfully by the net absorption-induced gravitational drag downward (ignoring wind resistance for the sake of this illustration).

What is interesting is that, while you are in free fall, if a rocket pack strapped to your back with its nose pointing *downward* is ignited, then your body tries to resist the additional downward force imposed by the rockets as your body reacts to the new rocket surge. That is because you are trying to impart an additional acceleration to your already-falling, already accelerating, body. The resistance that you feel is that of forced accelerative 'inertia' which is not at all the same as 'real' inertia. The rocket accelerates your body by acting as a surcharge through the straps of your rocket device and uses the bones of your body as a framework to push you faster downward. Yet, when you stepped off the building and fell, you would have felt no pressure through your body at all. This is the critical property of true gravitational acceleration, and it is why DOPA theory is so successful because it proposes only the use of true gravitational force on each individual atom of your body; it does not suppose that gravity is pushing one part of your body down like a rucksack on your shoulders.

110

So, how is inertia created by gravitational waves?

Figure 76 reminds us that at any point in space, including at the Earth's surface, there are omnidirectional, gravitational waves passing around and through the suspended weight.

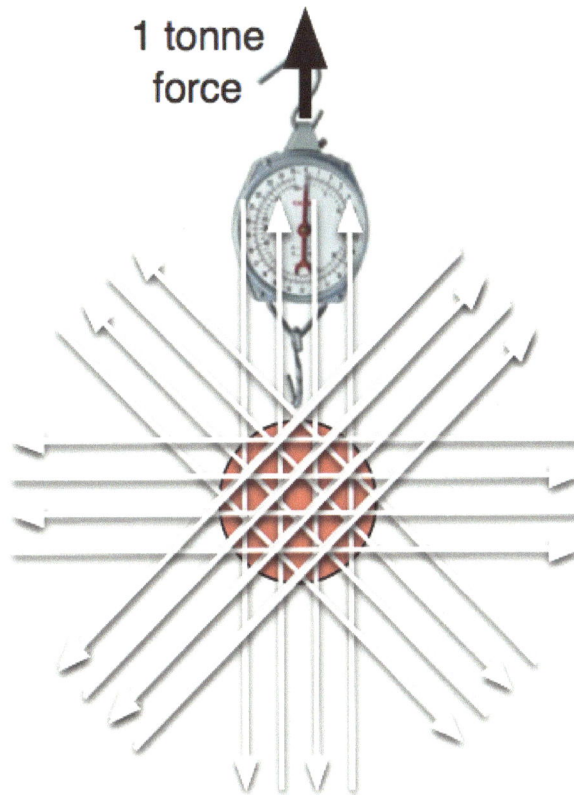

Fig 76. Omnidirectional gravitational waves pass around and through the suspended 1 tonne weight. In all directions except horizontally, there is a net downward vector force. Horizontal waves balance each other out in the force they create.

All of those waves except the horizontal ones will have a downward-vectored net force generated as a result of differential opposing partial absorption. The horizontal waves will be minimally depleted by their passage through the Earth's atmosphere, but equally so on all sides, and all horizontal waves will consequently be of the same amplitude and opposite in all horizontal directions, nullifying one another out to zero.

Figure 77 shows the same 1 tonne weight as in Fig 75, but now, a horizontal force is being applied such that the weight is being accelerated sideways.

In applying a left horizontal acceleration of 9.8 m/sec^2, we are creating an artificial, internal, partial gravitational wave absorption within the body of the 1 tonne mass. This is because we are accelerating the body through the otherwise-balanced horizontal waves.

In the balanced condition, the constituent atoms of the mass are dragged both to the left and the

111

right equally. This is analogous to the situation shown in Figure 75 where the vertical acceleration of the weight is being prevented by the supporting spring balance. In this case, each wave prevents the attempted acceleration trying to be caused by its opposing wave.

We accepted the statement given above, *"In short, weight is caused by a continuous state of **attempted** but prevented acceleration."*

INERTIA
Horizontally there is no force because
opposing wave amplitudes are equal

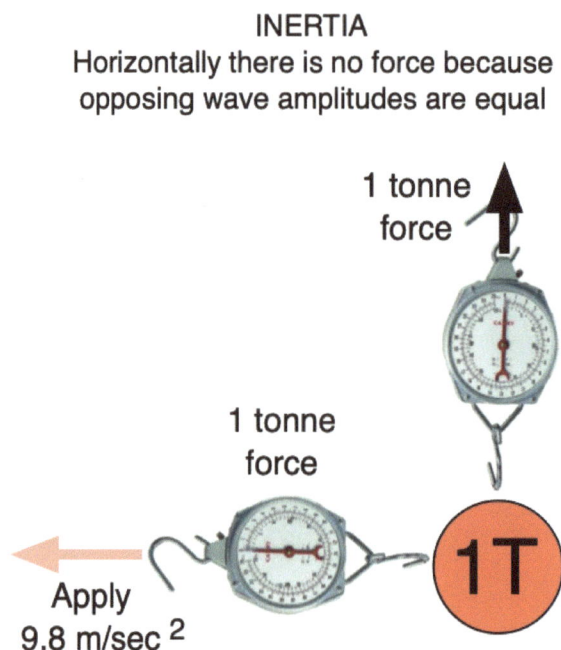

Fig 77. A brief acceleration of 9.8 m/sec^2 is applied towards the left of the suspended 1 tonne weight. It accelerates accordingly because it is not restrained laterally.

Thus, we can discount the balanced horizontal waves and consider what it is that stops us moving the 1 tonne weight horizontally with ease. If we think about it, we can realise that the resistance that we struggle against is the same as that of the strapped-on rocket, or the same as when we sit in our car seat and the driver puts her foot flat to the floor on the accelerator. In this case, when she presses the accelerator straight to the floor, the car lurches forward and keeps increasing in speed while, all the time, we are pressed into the back of our seat. In this circumstance, we are 'the weight', and the car is imposing a surcharge force on us. Most of us will be familiar with movie films of astronauts in space vehicles being subject to very rapid acceleration by the rocket motors. We see the astronauts being squashed most severely by the induced acceleration caused by the cockpit seats in which they are sitting.

From this, we can recognise that there are two 'types' of acceleration. One imposed by a gravitational field, and one imposed by an external pressure surcharge. When we are accelerated by a gravitational field, we do not even feel it unless we are restrained. We just step off the top of the Empire State building and let ourselves go. We accelerate downwards at a very fast rate, but we do not feel a thing except the air blowing past our faces. We are in 'free fall', just like an astronaut in orbit. We feel weightless. We do not feel any pressure pushing us downwards! That is because the gravitational waves are acting on every individual atom in our body; water, bones, blood, and flesh.

In contrast to that, when an external pressure accelerates us, we do feel it as the force of the external load is carried through our bone structure to force us in the direction of the load. Now, if we experience these two types of acceleration, so does any inanimate object such as a 1 tonne weight or a car.

So, we need to ask ourselves more clearly, why it is that when we have an external load applied to us by the car, our body resists it. Why does the car we try to push resist our pushing? Why does the swinging weight we try to move sideways resist us?

The author's proposal is that it is because we are trying to move the object through the gravitational field. If we move the object horizontally, we move it through horizontal-travelling gravitational waves which try to resist its movements.

Firstly, consider the special case where we apply a horizontal acceleration of 9.8 m/sec^2. We are moving the block horizontally in a preferential direction and upsetting the balance of the forces generated by the opposing horizontal waves, as shown in Figure 78. We are mixing the two types of acceleration. We are applying a force to the left to accelerate the object through its physical structure using the strength of its atomic bonds, and we are being resisted by true acceleration gravitational fields acting towards the right.

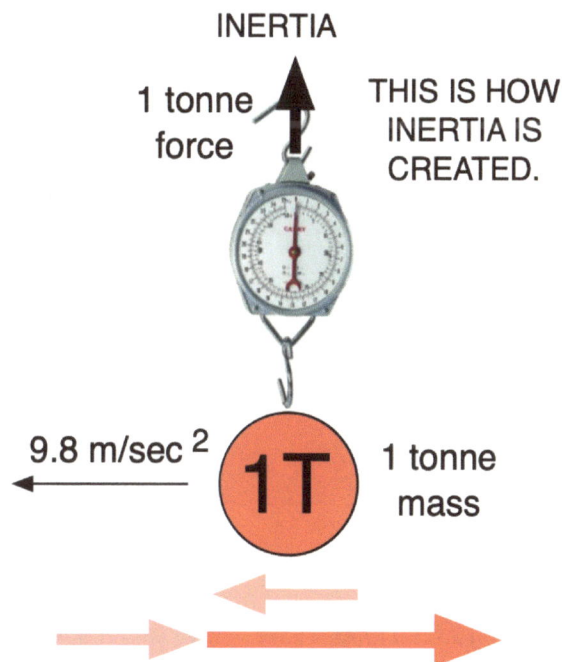

INERTIA

1 tonne force

THIS IS HOW INERTIA IS CREATED.

9.8 m/sec^2

1T

1 tonne mass

Fig 78. Left and right opposing forces are originally equal, so when a 9.8 m/sec^2 acceleration is applied to the left by an imposed force, an equal and opposing gravitational drag is created in the opposite direction (dark red arrow).

The essential mechanism to be explained is an INTERNAL mechanism deployed within the body of the object. This is how the absorption theory says all gravitational force is created—within matter. However, to examine how inertia is created we need to look at what is happening both within and outside the object concerned. In the case illustrated, it will take a continuously-applied horizontal force of 1 tonne to accelerate the 1,000 kg mass at 9.8 m/sec². Let us examine this in a little more detail.

First, look at Figure 71 and its accompanying arguments. Note that light impacts the surface at a fixed velocity. And remember that light comes in a great range of varying frequencies, even though its velocity is constant at 'C'. Now bear in mind that gravitational waves behave in the same way. They are not EMF waves, but they travel at the same velocity 'C' which, for the sake of clarity, we could call 'Cg'. There may not be any significant reason at present for allocating a different velocity notation from that of light, but future research and understanding may develop a need for a different notation. Thus, it seems sensible to allocate the notation 'Cg' at this early stage to the maximum velocity of the speed of gravitational waves.

Note that, since light can travel at C and still contain a broad spectrum of frequencies, and since light can experience Doppler effect shifts in wavelength, then there is no logical reason to expect gravitational waves to be any different.

Thus, we can expect gravitational waves to travel at Cg (where Cg = C) in a vacuum and travel slightly slower than Cg inside matter, as light does, and exist in a broad range of frequencies, and to experience doppler effects, and to be partially absorbed by matter as light is.

Figure 78 demonstrates the new proposal for how inertia is created, but it may be a good idea to look at his concept of wave-matter interaction to see why it works. To this end, Figure 79 represents a diagrammatic means of looking at the detailed interaction between gravitational waves and the atomic nuclei (or quantum fields) within matter.

Gravitational waves travelling towards the left

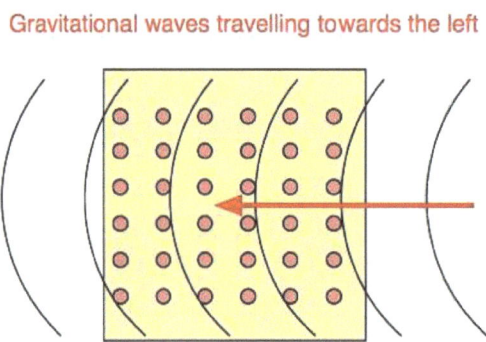

Fig 79. Atomic nuclei in a lattice within a solid matter cube, with gravitational waves progressing through from right to left.

The diagram of a cube may, in the first instance, be taken to represent half of the picture of the equal and opposite gravitational wave patterns affecting a particular cube of matter prior to a force being applied to move it. In reality, there is the exact opposite series of waves coming in from the left to the right that creates a balance. In effect, the left-travelling waves provide a resistance that prevents the right travelling waves from moving the object and vice versa.

This is directly analogous to the situation where the strong atomic forces within the atoms of the Earth's crust prevent us from travelling towards the centre of the Earth. That resistance creates our feeling of gravity because we are prevented from attaining the acceleration of 9.81 m/sec2 that the vertical gravitational waves are trying to force us to do.

In the same way, the horizontal gravitational forces opposing each other impart a sense of 'weight' to an object in the horizontal direction that we call inertia. In other words, although we do not feel it, because it is balanced, we are all subject to a horizontal weight from all sides. Fish living at the bottom of the Atlantic Ocean 2,000 metres down experience a pressure of 2,000 tonnes per square metre on their bodies but they do not notice it because it acts in all directions both inside and out.

The following are unrealistic numbers for the sake of illustration only.

Figure 79 represents a cube of matter containing 36 quadrillion (Q) atomic nuclei of density10g/cc being encountered by 3 gravitational waves per second with each imparting 10 units of drag per atom. Therefore, in this given cube of matter, there are is a total of 36Q x 3 x 10 = 1,080Q units of gravitational drag being imposed on the matter towards the left. At the same time, there are exactly 1,080Q drag units being imposed to the right. They balance out.

Now, we have stated and explained why the gravitational waves would strike both sides of an object at the same speed Cg irrespective of the motion of the object, which means irrespective of the object's prior velocity (zero in its own initial frame of reference) or its post-acceleration, moving, frame of reference. But, although each wave must strike the surface of the object at exactly Cg, there is no reason why the object cannot 'gather up' additional waves in an internal Doppler effect, thus increasing their frequency internally within the object while maintaining the Cg velocity constant. (Note that this argument holds true whether or not Cg is slightly reduced when travelling through matter because it applies to all waves travelling through in both directions.) In effect, you change the frequency with which waves pass through the object, but not the velocity of the waves.

So, let us say that in the case of the cube of matter illustrated in Figure 79, there may be within the cube 1,080Q drag units in balance at the instant before the lateral accelerating force is applied, and let us say that the acceleration applied from right to left changes the internal frequency from 3 per second to 4 per second travelling from left to right. Note that we do not change the velocity of the gravitational waves, merely their frequency inside the body of the weight as we intercept and scoop up more per second.

Increasing the frequency of waves travelling through the cube laterally by one third will mean that each atom encounters four per second when previously it encountered three. So, there is a build-up of drag on each atom in the ratio of 1/3.

This is an additional one-third of the original lateral force acting against our pressure. In that case, the pre-existing balance in the wave frequencies is changed. Now, the atomic nuclei experience an additional 36Q x 1 x 10 = 360Q units of drag acting from left to right. That is resisting our imposed acceleration from right to left.

115

Now there is an imbalance of forces within the cube that is identical in nature to that which we feel when we are weighed down onto the surface of the Earth by the planet's natural gravitational imbalance. In nature, the waves move through us, but in this case, we are moving the cube through the waves. The outcome is the same.

We have created this new imbalance, but nonetheless, it is as real as that of any other imbalance and the numerical conditions match also. If we apply a lateral force of 1 tonne, to a mass of 1,000 kg, we shall obtain a lateral acceleration of 9.81 m/sec^2.

This imbalance (the increase in the frequency of waves encountered) will only exist while we continue to apply the lateral force; as soon as we stop accelerating the cube laterally, the resistance will stop and the cube will continue to travel under its own momentum.

The reason why this argument works when compared to the argument concerning a planet moving in orbit in space (See Figure 71), is that the planet in orbit is not being accelerated, and secondly, when considering an accelerated body, the relative wave frequency is being generated within the body of the mass being accelerated and is thus being considered within the body's own internal frame of reference. The argument for an orbital body was that light impinges on the body at a constant speed irrespective of the body's velocity because the light travels in the same frame of reference as that of the planet in orbit. In the case of the 1 tonne weight we discussed above, once the waves are inside the body, they are within its own frame of reference.

This mechanism explains how inertia is created, why inertia is created by gravitational waves, and why inertia is exactly proportional to the acceleration and the mass of the object being accelerated.

Because gravitational waves impinge on both the left and right surfaces of the body at the same velocity (analogous to light), owing to relativity, it is immaterial as to whether the cube is moving in a given reference frame or not. The gravity waves arrive at the surface at the velocity of Cg — the equivalent of the speed of light. However, that does not prevent them from exhibiting doppler effects where their frequency and wavelength can change within matter as they arrive at the surface of an accelerating object. Note, not just a moving object, but an *accelerating* object which is transferring from one frame of reference to another. Acceleration can cause an internal doppler effect to take place as illustrated above.

When applying a right-to-left horizontal acceleration, we were causing one extra left-to-right waves to pass through additional number of atoms in one second as if the wave was an isolated wave travelling to the right. It is the resistance of that one extra wave per second that we feel when we try to accelerate the weight to the left.

Remember to bear in mind that the opposing waves still grab and oppose the same number of atoms per unit time as before, so they are not relevant to our considerations.

Note that it is only during the period of acceleration that the resistance of inertia is generated. If the accelerating force is suddenly removed, then the object moves at a new constant velocity in its new frame of reference, and the impinging waves become instantaneously balanced again. In other words, to experience inertia, the body has to keep scooping up extra waves, so that the resistance

116

can be maintained.

Instead of the waves dragging themselves against the atoms, as happens naturally, *we* impose an acceleration that drags the atoms against the waves, thus creating an identical resistance.

Wikipedia says, among other things, *"No physical difference has been found between gravitational and inertial mass in a given inertial frame. In experimental measurements, the two always agree within the margin of error for the experiment. Einstein used the fact that gravitational and inertial mass were equal to begin his general theory of relativity, in which he postulated that gravitational mass was the same as inertial mass ..."*

https://en.wikipedia.org/wiki/Inertia

The author concurs with Einstein, and proposes that there is only gravitational mass, being the same as inertial mass, which is created by the mechanism proposed in DOPA theory.

5.2   The proposition that inertia and momentum are the identical same thing caused by the same mechanism.

The author explains below how DOPA theory demonstrates that momentum is simply inertia viewed from a different frame of reference.

Consider a large diameter black-painted tube containing two magnetically-levitated (maglev) vehicles. Both are stationary in their initial frame of reference on a horizontal track in a vacuum. A colleague sits in one and you sit in the other. Both are fitted with small rockets at both ends.

You sit on your vehicle and fire one of your small rocket engines which accelerates your vehicle and you to 5 km/hr. You certainly feel the inertia of your body trying to hold you back as the rocket fires, but you hang on and after a few seconds the rocket stops firing. As far as your colleague is concerned you are moving away relative to his frame of reference. But, because you cannot see any detail on the walls of the tunnel and because the maglev is totally smooth so that you don't feel any vibration, as far as you are concerned, sitting in your vehicle you are now stationary and according to your new frame of reference your colleague appears to be travelling in the opposite direction. Yet you know that he feels stationary in his frame of reference.

Now—and this is the very interesting thing—if you fire the rocket in the opposite direction, your colleague sees this action as a 'breaking' device slowing your vehicle down relative to him. But, to you, now considering yourself to be stationary while your colleague moves away, the rocket firing in the opposite direction is accelerating you back towards your colleague, and you feel the acceleration force in the opposite direction to your original feeling.

You can stop your rocket during its second burn, to find yourself stationary with regard to your colleague, although you have moved a distance away from him. You are now back in his frame of reference. The critical thing is that he saw your actions as a) overcoming inertia under acceleration away from him and b) deceleration to overcome your new momentum until you became stable again. You, on the other hand, saw them as acceleration in both cases, *and in both cases, you saw it as overcoming inertia.* **You overcame inertia to create a second frame of reference and then you overcame inertia to move back to your first frame of reference.**

Furthermore, although you have moved away from your colleague, if you now repeat the exercise in the opposite direction, you can overcome inertia twice more to bring you back to the exact same location as where you started. To precisely the original frame of reference. Your colleague will see these second steps as exactly the same as steps a) and b) above. He will see you overcoming inertia to accelerate back towards him and then he will see you overcoming your new momentum to become stationary exactly where he is and where you started. He sees it as inertia/momentum/inertia/momentum. You see it as inertia/inertia/inertia/inertia. You see it this way because you changed your frame of reference twice.

In all of these changing frames of reference, once you stopped accelerating, you were—as far as incident gravitational waves are concerned—stationary. As soon as you stop accelerating, gravitational waves start to pass through you from all sides at the standard Cg velocity. Thus, you can tap into the Roberts mechanism as many times as you like to overcome inertia again and again. The fact that your colleague views it differently as inertia and momentum does not alter the fact that in each of your changed frames of reference, you are stationary vis a vis the essential gravitational wave mechanism.

Thus, your experience has shown us that **inertia and momentum are the same things exactly, just viewed from different frames of reference.**

This also means that the atomic drag explanation for the sensation of inertia resistance provided by DOPA theory also provides the explanation for overcoming both inertia and momentum.

Note what Wikipedia says about inertia—*that it has so far eluded explanation to the point where some scientists believe that there is actually is no mechanism involved.*

Those scientists simply resort to the ancient Greek alternative of saying that the property of inertia simply exists without a mechanism. That is, to the author, an untenable retrogression, now made redundant by the author's reasoned gravitational mechanism.

5.3   Absorption theory has provided the explanation for why an independent piece of laboratory work shows how gravitational force is affected by light.

Laboratory research work undertaken in 2011 and 2015 by Professor Louis Rancourt showed that light appeared to vary the gravitational force exerted on a weight attached to a delicate torsion balance when the light was shone *both above and below* the weight. This has already been mentioned, in passing, in Sections 4.7 and 4.12. Note that the light was not shone above and below the torsion balance itself, but just above and below the weight itself, as illustrated in the following diagrams.

[Louis Rancourt, "The Effect of Light on Gravitation Attraction", Physics Essays, 24(4), 557-561 (2011)]

[Louis Rancourt & P.J. Tattersall, "Further Experiments Demonstrating the Effect of Light on Gravitation" Applied Physics Research Vol 7, No 4 (2015)]

Consider the following diagram. Figure 80 shows what Rancourt and any other scientist at that time might consider would happen if they inserted some proposed 'gravitational barrier' between an object and the Earth. They expected that, if a weight on a sensitive torsion balance *were connected by gravity to the Earth*, then if a barrier of light was going to affect that connection it would have to be placed between the weight and the Earth. If it were to be set up above the weight, it would have no effect. That was a reasonable expectation. They expected that when the light barrier was placed between the weight and the Earth, the reading of the value of gravitational attraction would be reduced as shown in the diagram of Figure 80.

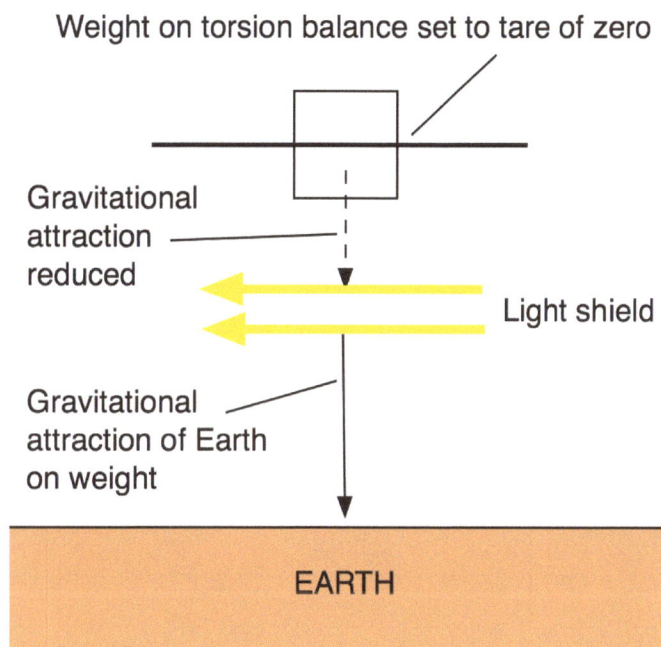

Fig 80. Rancourt's apparatus. He expected that a light shield would reduce the gravitational pull on the weight on the torsion balance.

Amazingly, the opposite happened. When the light barrier was placed beneath the weight, the balance's reading of the object's weight *increased*. Then, even more amazingly, when the light barrier was placed above the object, there was recorded an apparent *decrease* in the object's weight.

These results seemed quite nonsensical to them, as, indeed they did to the author until he considered the results in the context of his own new theory for the creation of gravitational force. Figure 81 represents the explanation of the gravitational setup that differential opposing partial absorption theory provided, which is entirely different from the expected Newtonian setup in Figure 80.

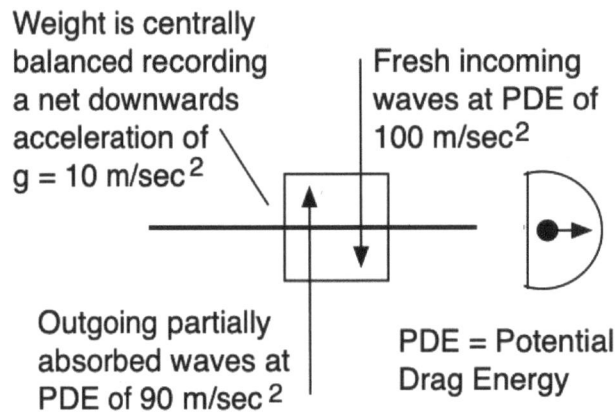

Fig 81. Rancourt apparatus with weight on balance, and with zero tare deflection shown on the scale. Values shown are fictitious for purposes of illustration only

In Figure 81, we see a diagram representing Rancourt's apparatus set up ready for the experimental work to begin, containing the author's interpretative illustration of what is happening according to DOPA absorption theory. (For example, to simplify the numbers, and to emphasise their imaginative nature, the net acceleration of gravity has been approximated from 9.81 m/sec² to 10 m/sec².) The diagrammatic indicator at the right-hand side shows a level, zero tare deflection.

According to the Newtonian concept of gravitation, the Earth directly 'attracts' the load on the balance with an attempted acceleration of 9.81 m/sec² as shown in Figure 81—that was common knowledge at the time. Or, alternatively, the weight was trying to move downwards because the Earth was distorting spacetime. Exactly what the light would do to change spacetime if that were real, it is impossible to say.

Under **opposing** absorption theory, however, the load being measured by the torsion balance is applied in a fundamentally different way. The balance weight is obtained, according to the DOPA analysis of the situation, by the weight on the torsion apparatus being subject to an incoming, downward, net gravitational accelerative drag of, say, approximately, 100 m/sec² and an outgoing, upward, degraded, net gravitational accelerative drag of, say approximately, 90 m/sec². These counter-balancing forces result in an overall, net, downward gravitational acceleration of approximately 10 m/sec² (9.81 m/sec² in the case of Earth).

The reader will recall that the actual value of the downward and upward potential drag strengths are unknown except that they must be considerably higher than these convenient figures and their difference is, in our case, 9.81 m/sec². The reason we know that the values of the downward and the upward forces are extremely large is that the maximum, incoming, figure must be sufficiently large to control the geometry and behaviour of the very large structures we observe in the universe, including massive stars, binary star pairs, and even galaxies themselves. For the sake of this exercise, though, simple, very small numbers have been adopted. Their values are immaterial to the conclusions drawn, which depend only on the difference between the two figures chosen and the assumption that these are real waves travelling along real vectors.

The difference in the two interpretations of how gravitational force is developed (Figs 80

Newtonian and 81 Robertsian) is critical to being able to interpret the experimental results obtained by Rancourt and Tattersall.

Having set up the apparatus in balance, the light 'shield' was first shone below the weight (See Figure 82)— potentially, so the scientists thought, to partially cut off or otherwise interfere with the gravitational attraction between the measured object and the pull of the Earth. The researchers were surprised when the balance showed a reaction (as they had hoped), but the balance was pulled downward! This could only mean that, somehow, the light shield was enhancing the effect of gravity and pulling the balance down even more strongly. Some kind of 'amplification' effect, perhaps? They could not explain this rationally at all. That was the first surprise.

Then, simply for the sake of interest, and not expecting any result at all, the light 'shield' was shone above the weight. The researchers expected that nothing would happen and were doubly shocked to find that a) there was an effect where none should exist (because the light no longer acted between the Earth and the weight), and that b) the apparatus mechanism now moved upward! This meant that, somehow, the light shield was reducing the effect of gravitation between the weight and the Earth, and making the scales read lighter.

In both tests, the deflections of the weight were very tiny but recordable.

These findings were inexplicable in terms of both their operational mechanisms and the direction in which they appeared to operate. The puzzling points, forced on the researchers, were:

i) What physical mechanism could possibly be involved in light affecting the force of gravity between the Earth and another object?

ii) What physical mechanism could possibly be involved in both enhancing and inhibiting the force of gravity between two objects?

iii) How could any physical mechanism operate above the level of the apparatus when the force of gravity was, supposedly, exerting itself between the apparatus and the centre of the Earth below?

Unfortunately, neither Rancourt nor Tattersall was able to provide any explanation for the experimental results. Nonetheless, they very courageously, in the true spirit of experimental science, chose to publish their results *and* the fact that they could not explain them. (Professor Rancourt has asked the author to make clear that, although the results were confusing, he concluded that the results did show that light affects gravity and he and Tattersall came to the further conclusion that gravity must be a pushing force coming from all directions in space. He believes that the Earth can block some of the pushing force and "the net sum of all the vectors of these forces is a downward vector pointing to the ground. That would cause the weight force.")

The important thing for the reader to recognise in that statement from Rancourt is where he says, "...he and Tattersall came to the conclusion that gravity must be a pushing force coming from all directions in space." This is, unfortunately, no better than a hypothetical suggestion because it contains no scientific support. It merely states that they consider gravity to be a pushing force." Such statements do not take us forward in any understanding of what gravity is. Simply saying that it is a pushing force does not take us one step further than De Sage in 1748 who said the same thing, with the same lack of scientific support. Rancourt's experiments produced results that he could not explain. He says that their conclusion was that gravity must be a pushing force, but, unlike DOPA theory, he could not explain why a pushing force would produce the output results of the experiments. Differential opposing partial absorption theory, on the other hand, can and did—even though it was some years after the research had been published.

This 'pressure' concept was described in the book "Gravity-why what goes up must come down." by Brian Clegg. Published in 2012 by Duckworth Overlook, 90-93 Cowcross Street, London EC1M 6BF.

In his book, Brian Clegg wrote:

"Like Newton's own, these theories relied on some sort of medium existing throughout space, whether that same "ether" that was thought to carry the waves of light or some other substance.

"The longest lasting of the theories was originally developed by the Swiss mathematician Nicolas Fatio de Duillier, but later (and possibly independently) conceived by his countryman, physicist Georges-Louis Le Sage. Variants on this theory would be produced all the way up to the beginning of the twentieth century, most notably by William Thomson (Lord Kelvin), though Le Sage published his ideas in 1748.

"Most ether-based ideas considered the ether to be a continuous fluid of some sort, but the de Duillier theory gave the ether a Newtonian twist. Newton believed that light was made up of particles, or corpuscles as he called them. In the mechanical theory of gravitation that de Duillier and Le Sage proposed, the ether itself had a corpuscular structure. These particles weren't static, but instead flew around in all directions, smashing into solid objects. Just as we now know that air molecules slamming into objects causes air pressure, it was thought that these gravitational corpuscles would influence the bodies with which they collided.

"If a single object was isolated in space, it would not move as a result of the pressure from the corpuscles because they came at it from all directions in equal quantities. But think what would happen if two bodies occupied space relatively near each other. Each body would be shaded from corpuscles traveling from the direction of the other, just as the Moon shades the Sun and causes an eclipse. There would be fewer corpuscles hitting the bodies on the sides facing the other. The result? The two bodies would feel a net pressure towards each other. They would be attracted. What is particularly neat about this model is that it automatically generates the inverse square law that is so central to Newton's workings. But over the years it was proposed, many were doubtful about it, because the solution seemed as bad as the problem. This was a mechanical approach that did away with action at a distance, but it required the universe to be full of streams of invisible undetectable corpuscles that had a combined effect that was powerful enough to keep the planets in their courses.

"Although Newton never came up with the formula involving G, his work on gravitation was to have a huge impact— yet there was an issue that even he would acknowledge. He had no clue as to how gravity did its stuff."

And—like Newton—Professor Rancourt, in his written submission to the author, could still only say, *"that light somehow interferes with the force of gravity."* DOPA theory has provided an answer to the word, "somehow".

Note from the author.
The concept of a pushing force is the one that I described when I wrote my first paper in 1978. I abandoned this approach subsequently because it became apparent to me that such a 'pushing' force does not fit in

with the internal stresses experienced in structures created and used by civil engineers. All objects subjected to gravitational acceleration are stressed internally, not as a surcharge load—which would be imparted by surface pressure as proposed by Le Sage in 1748, Roberts in 1978, and Rancourt in 2015. This theory's internal drag mechanism, on the other hand, re-creates the gravitational acceleration forces experienced in self-weighted structures perfectly. This has been covered in more detail in Section 3.10 of this book.

(Professor Rancourt has also asked the author to refer the reader to Michelin's paper of 2013 (unspecified), and to the paper of Libor Newman who independently discovered the same effect in Prague, a few years after Rancourt's work and published his findings in PHYSICS ESSAYS 30, 2 (2017).)

In December 2017, having developed his own theory for the creation of gravitational waves, the author became aware of the research papers of Rancourt and Tattersall, and read their papers with their lack of explanation. He realised, as he read, that his own theory *could* explain the results and answer all three of the questions i) to iii) above, as follows:

Figure 82 shows the reading of the apparatus when a strong shield of coherent laser light was shone below the apparatus. The light layer was not connected to the apparatus in any way but was positioned below the weight in order to intervene between it and the Earth. This was logical, and this setup was established to see whether the light would interfere by being placed within the line of attraction being exerted by the Earth on the apparatus.

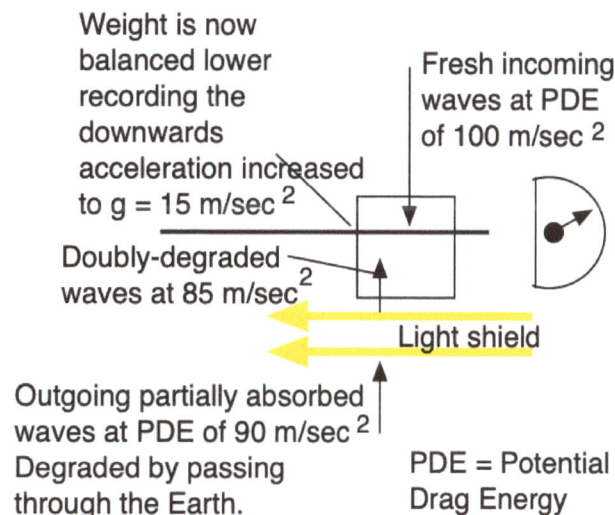

Weight is now balanced lower recording the downwards acceleration increased to $g = 15$ m/sec^2

Fresh incoming waves at PDE of 100 m/sec^2

Doubly-degraded waves at 85 m/sec^2

Light shield

Outgoing partially absorbed waves at PDE of 90 m/sec^2 Degraded by passing through the Earth.

PDE = Potential Drag Energy

Fig 82. Rancourt apparatus with light shield interposed below the weight between it and the Earth. DOPA explanation.

Under absorption theory, there are two sets of gravitational waves resulting in the formation of the null tare balance reading of the apparatus upon original setup (see Figure 81). When the light is shone below the apparatus, the only gravitational waves relevant are those coming up out of the Earth because they are interfered with *before* they reach the weight to impose their upward drag on it. The light actually does interfere with the downward waves, but those have already interacted with the apparatus before being interfered with. Thus, the light reduces the up-going waves before reaching the apparatus and not the down-going ones. The reader can observe, from Figure 81, that a reduction from 90 to 85 in the upgoing wave will increase the overall effect on gravity by increasing the net balance to 15 m/sec^2 downward. The small meter at the right-hand side of the diagram shows that notional increase in gravitational force.

Differential opposing partial absorption theory, therefore, explains why placing the light between the apparatus and the Earth could actually *increase* the apparent force of gravity affecting the apparatus.

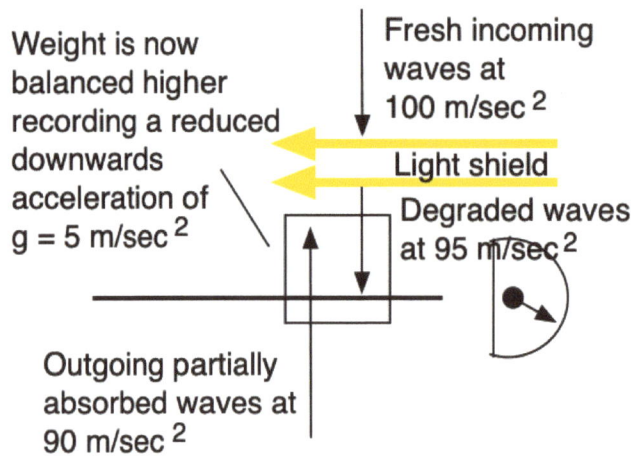

Fig 83. Rancourt apparatus with light shield placed above the weight—not between it and the Earth. DOPA explanation.

Figure 83 shows the reading of the apparatus when a strong shield of coherent laser light is shone above the apparatus. Logically, according to the traditional concept of gravity, this should have had no effect on the apparent weight of the reading mechanism in the apparatus. However, it did.

The light layer was not connected to the apparatus in any way but was positioned above the weight in order to demonstrate no effect, thus proving that the effect of gravity is exerted between the weight and the Earth. It would prove this by having no effect on the null reading. Unfortunately for the researchers, an effect was measured where none should have existed. When the light was switched on, the readings on the torsion became lighter and the weight, apparently, weighed less.

Under DOPA theory, as shown in Figure 83, there are two sets of gravitational waves to be considered, and it is the down-going one with which the light now interferes. When the light is shone above the apparatus, the only gravitational waves relevant are those coming down towards the Earth because they are interfered with *before* they reach the apparatus to impose their downward drag on it. The light does interfere with the upward waves, but those have already interacted with the apparatus before being interfered with. Thus, the light reduces the down-going waves before reaching the apparatus and not the up-going ones. The reader can observe, from Figure 83, that a reduction from 100 to 95 in the down-going waves will decrease the overall effect on gravity by decreasing the net balance to 5 m/sec^2 downward. The small meter at the right-hand side of the diagram shows that notional decrease in gravitational force.

So, we can now see how absorption theory has unravelled this puzzle.

124

Absorption theory explains why placing the light above the apparatus, on its opposite side from the Earth, could actually decrease the apparent force of gravity affecting the apparatus.

The significance attached to this work by the author is that since Professor Rancourt says Michelin and Newman have corroborated his experimental results, their work provides some additional support for DOPA theory since no other theory could explain Rancourt's results. The reason for that is gravity has never before been thought of as being the difference between two sets of forces.

5.4   The development of DOPA theory has prompted the author to outline some of the consequential outcomes on the development of gravitational force within a planet or star.

The theory that gravitational waves create the gravity force that holds together any cosmic body leads us to have to recognise the following principles.

Firstly, note that Figures 61 and 62 represent the current, tentative, form of the PREM (Preliminary Reference Earth Model) for the variation of gravitational acceleration with depth from the surface value of approximately 9.81 m/sec^2. Also, note that, as shown in Figure 62, it would seem that these results are calculated using Newton's equation, operating under the same principle used to calculate the supporting shell theory evidence.

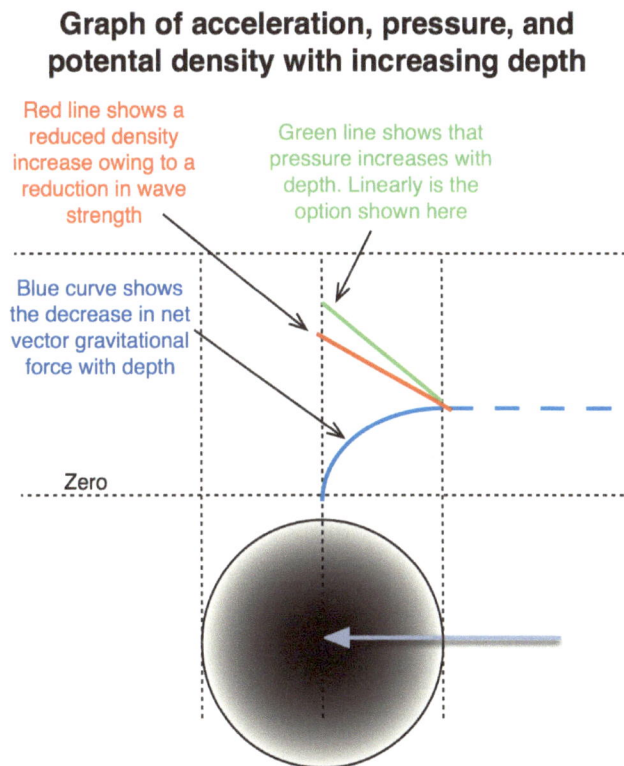

Graph of acceleration, pressure, and potential density with increasing depth

Red line shows a reduced density increase owing to a reduction in wave strength

Green line shows that pressure increases with depth. Linearly is the option shown here

Blue curve shows the decrease in net vector gravitational force with depth

Zero

Fig 84. Simplified, diagram indicating qualitative relationship between gravitational wave vector force, theoretical pressure response and actual pressure response in a planet according to DOPA absorption theory.

125

Secondly, the various examples and arguments put forward in this book explain that DOPA theory does not agree with the Preliminary Reference Earth Model (PREM) which proposes that gravitational force will increase from the surface in a planet or star with a high-density core. The partial absorption principles of the theory mean that, whether taking a uniform density body, a body whose density increases gradually with depth, or a body with a denser core, the net force of gravity would only decrease steadily towards the centre. The author considers that the curve of the gravitational plot will reflect the internal structure of the planet or star being considered, in terms of its varying density from surface to centre.

In order not to go over the whole PREM ground again, the author wishes simply to make some brief observations as to what happens within the Earth primarily as the result of considering the partial absorption concept. Interestingly, much of this applies to the PREM construction and applies to other planets and standard non-shell stars as well.

It is a very important part of DOPA theory that gravitational waves are partially absorbed when they pass through matter—albeit minutely. The denser the matter, the greater the amount of absorption because the greater the number and weight of the atomic nuclei encountered. Having said that, it is proposed that any gravitational wave can pass through a normal star and pass out again, hardly altered.

Even with a standard black shell star, it appears that pristine gravitational waves can pass in, through and out with ease, being only slightly depleted. Larger black shell stars, however, are likely to be sufficiently dense and massive as to absorb a significant percentage of the amplitude of a passing-through gravitational wave.

Figure 84 looks at some qualitative properties of a planet with regard to the partial absorption of gravitational waves and the consequent effect on the internal force exerted and the body's density. In Figure 84, a qualitative blue-coloured curve can be seen indicating the Roberts decrease in gravitational net vector force with depth from surface to centre of a linear-density-increase planet. This book has already explained, in and around Figure 24, that at the centre of a planet or star, the net vector of all incident omnidirectional gravitational waves is zero. The same applies to any star.

However, that does not mean that the wave force itself has fallen to zero. Of course not. It is only the summed vector that is zero (see Figure 24). The gravitational waves have retained the vast majority of their full strength. If they had retained their full strength, then the increase in actual pressure would have followed the green line in Figure 84, assuming a constant density. However, there is the matter of gravitational force loss as a result of the energy lost in imparting a tiny gravitational drag to each atom encountered in the planet. In the case of the Earth, the total loss at the planet's surface is approximately 9.81 m/sec^2.

Consequently, pressure increase and consequent density reduction would actually follow a slightly reduced line as indicated by the red line in Figure 84.

For ordinary planets and stars, the total real loss in gravitational force is relatively small. Considering the counteracting forces involved and the outcome of surface gravity on the Earth, for example, the total real loss of wave amplitude is minuscule. Nonetheless, it is considered necessary to mention these small points.

Equally, it is necessary to mention the following important point. When very massive black shell stars are considered, where the gravitational wave absorption losses are significantly large, then there is experienced a correspondingly significant loss of the driving force creating pressure with depth. In the case of the Figure 84 example, there is little difference between the theoretical increase in pressure and the actually realised one. However, with very massive black shell stars, it is proposed in absorption theory, that pressure no longer increases towards the centre of the star in the conventional way. Whereas the pressure increases inwards, at the same time the gravitational waves are becoming steadily weaker. It is reasonable to suppose that, as the waves become weaker with depth towards the centre, *the rate of pressure gain* slows. Ultimately, this process continues until the incoming gravitational waves become *totally absorbed* and can no longer, subsequently, create any accelerative drag in the atoms of the star. This creates a "gravitational shell star".

The gravitational shell star

As a result of total absorption of gravitational waves, no further drag can be applied inside the gravity-free core, and **a horizon shell develops where there is no longer any increase in pressure with depth at all**. This is an entirely new concept that results directly from the Roberts DOPA theory. It is described in detail below, but the important thing is that below that shell depth, there is no driving force trying to increase the density of the core. *This results in a whole new series of theoretical possibilities that have never been considered before.*

No one, to the best of the author's knowledge, has ever suggested before that there could be a maximum pressure that can be created by nature. To explore this in more detail, consider Figure 85.

Absorption Stage 1

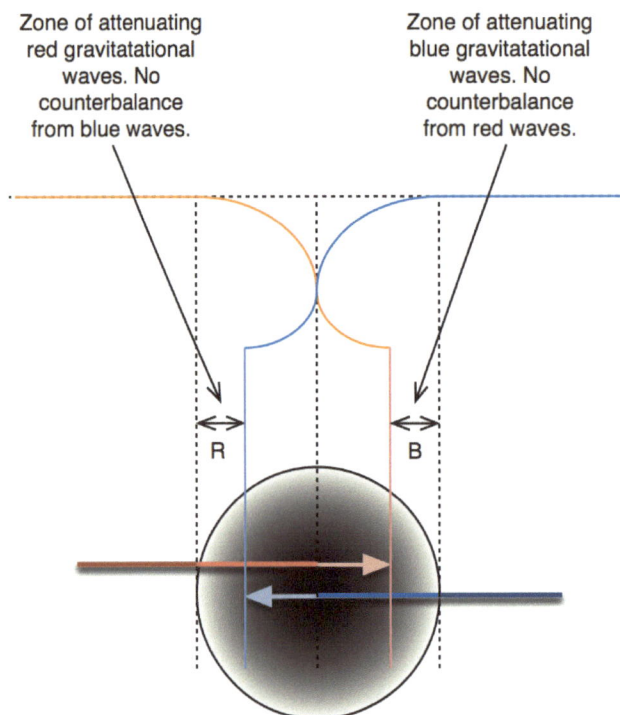

Fig 85. Stage 1. Total absorption of gravitational waves takes place before the waves can exit the star. The waves pass the star centre before becoming totally absorbed and disappearing.

In Figure 85, we can see that both the red and blue wave sets are partially absorbed in the usual way as they enter into the star and pass towards the centre. However, in the case of this very dense star, they have suffered sufficient absorption that, by the time they have travelled from the centre halfway to the surface of the star, they have become totally absorbed and disappear.

Thus, in zones R and B there exist only inward-vectored, partially-absorbed, gravitational waves without counter-balance from any outgoing waves. Zones R and B represent an outer shell of the star in which there exists an almost-un-resisted inward force of very high level, which, in turn, subjects the inner zones of the star to increased pressure beyond the norm shown in Figure 84. This could be a potential candidate mechanism for the initiation of core collapse.

Figure 86 shows this same state of affairs in a diagrammatic form rather than as a graph. The arrows show how the omnidirectional incoming waves travel through the centre but become entirely absorbed before they can exit the star.

128

Absorption Stage 1

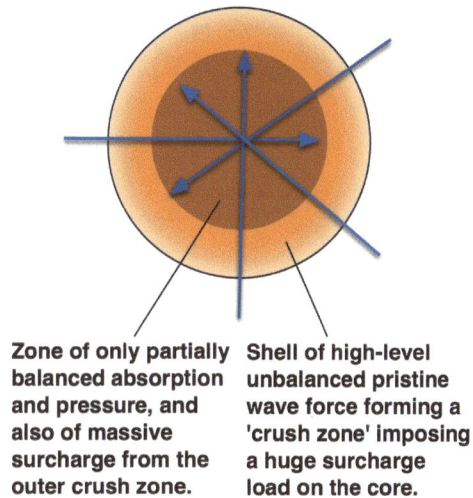

Zone of only partially balanced absorption and pressure, and also of massive surcharge from the outer crush zone.

Shell of high-level unbalanced pristine wave force forming a 'crush zone' imposing a huge surcharge load on the core.

Fig 86. Stage 1. Total absorption of gravitational waves takes place before the waves can exit the star. The waves pass the star centre before becoming totally absorbed and disappearing.

Figure 85 shows that when a black shell star increases sufficiently in mass, it develops an internal "*gravitational* shell", through which <u>outgoing</u> waves do not pass owing to their having become totally absorbed. At this stage, it is now called a Stage 1 U_{max} star, where the notation 'U_{max}' represents the maximum pristine potential gravitational wave force generated in any particular part of the universe.

As a Stage 1 star continues to become bigger and more massive, the internal gravitational shell shrinks inwards and eventually reaches the centre. In terms of Figure 85, zones R and B become wider until they eventually meet at the centre. At this precise juncture, the pressure being exerted on the centre of the star reaches its maximum level. Subsequent to this, with the gravitational wave energy being totally absorbed, they have no more drag energy to give and, therefore, below the absorption shell, there can be no increase in pressure. This is a unique concept.

This process produces the end of Stage 1 and the start of Stage 2, being a star in which all incoming gravitational waves are totally absorbed by the time they reach the centre and in which there is a core of constant pressure. The level of gravitational load is extremely high, trying to crush the core region of the star even more, but it never increases any more. Thus, the liquid core experiences a constant pressure throughout and does so as the star continues to grow by accretion. That also means that the core retains a constant density throughout. The consequences of this are addressed in Section 5.5.

Stage 2 progresses, as shown in Figure 87, such that the gravitational waves become increasingly more rapidly absorbed. In both Figures 87 and 88, pristine gravitational waves enter the planet and are totally absorbed by the time they get less than half way to the centre of the planet.

In the case of any smaller star, it is speculative as to where one can say precisely where its surface exists. It is possible that, in the case of very massive black shell stars, the high gravitational forces create a less-turbulent surface than on conventional non-shell stars.

Absorption Stage 2

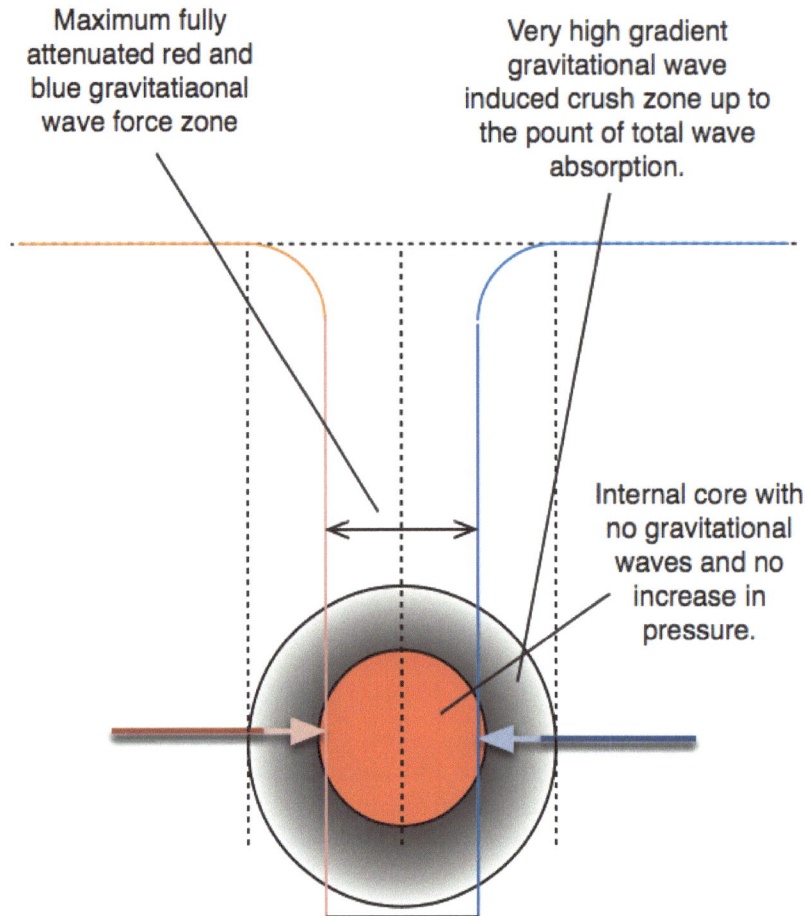

Maximum fully attenuated red and blue gravitatiaonal wave force zone

Very high gradient gravitational wave induced crush zone up to the pount of total wave absorption.

Internal core with no gravitational waves and no increase in pressure.

Fig 87. Stage 2. Total absorption of gravitational waves takes place before the waves reach the centre of the star. Red core is constant pressure and constant density. Red and blue graphs fall to zero.

Figure 87 graphically represents the development of the Gravitational Stage 2 star, in which incoming gravitational waves are completely absorbed while on their way down towards the centre of the star, forming the Stage 2 "*gravitational* shell". This creates a new, heretofore-unknown, spherical horizon enclosing a spherical core devoid of gravitational waves. Thus, an internal gravity-wave-free zone is surrounded by the near-maximum gravitational force pressure that nature can provide—full, pristine, virtually-undiluted gravitational wave pressure.

Ultimately, although strictly unlikely, the Stage 2 gravitational shell could reach towards the star's surface, whereby the pristine incoming gravitational waves are absorbed so quickly down from the surface that, in effect, they exert a pressure on the whole of the star which becomes almost gravity-free within.

130

At this stage, imagination runs riot. An entire super-star held in by pristine gravitational waves exerting the maximum force and pressure that can be created by nature. How unstable such a structure would be is hard to imagine. One could speculate that such a situation would present an ideal mechanism for the star to turn supernova as shown in Figure 68, returning the super-shell star to a smaller black shell state, to start its growth over again, entering a cycle of change.

An alternative scenario, it seems to the author, is that the internal pressure induced within the star at the Stage 2 event horizon would rise to match the maximum possible pristine pressure level and, for some presently-unseen reason, the super-shell star would stabilise about the stage shown in Figure 88. Also, it follows, during this stage there would be the novel condition of the star's core having no inertia because of having no gravitation. This is explored below.

5.5   The proposition that there may be a maximum force that nature can exert, a maximum pressure that nature can create, a maximum density of material that nature can create, thus preventing the formation of a singularity, and a core without gravity or inertia owing to the lack of the presence of gravitational waves within it.

These conditions follow on from the propositions in Section 5.4

It is interesting that the maximum pristine gravitational potential force figure must be so high that it can control the formation, shape, and gravitational behaviour of a planet, a star, a solar system, a giant star, a black shell star, and even influence an entire galaxy.

The first fascinating new concept that derives from this situation is illustrated in Figure 88, that the core of the star is now free of any gravitational waves at all. Hence, within the core, there is no gravitational force at all and pressure is uniform in all directions and cannot increase with depth towards the centre of the star because there are no gravitational waves to drive any pressure increase there. A special form of hydrostatic pressure exists within this core zone. As the star grows in size, the pressure within the core will increase uniformly whether it is in a solid or liquid state.

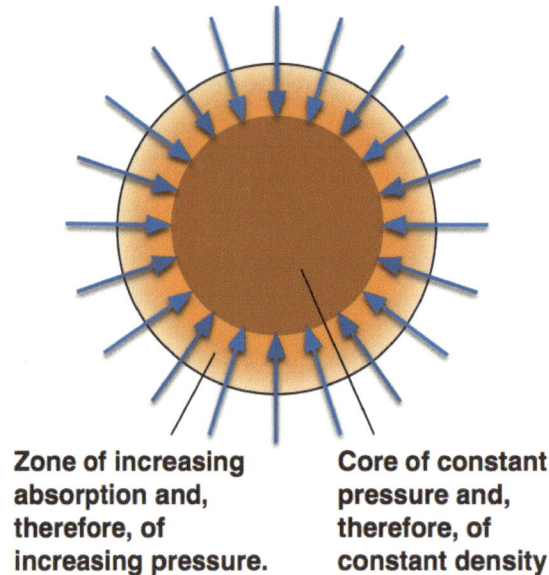

Zone of increasing absorption and, therefore, of increasing pressure.

Core of constant pressure and, therefore, of constant density

Fig 88. Stage 2. Total absorption of gravitational waves takes place before the waves reach the centre of the star creating a constant pressure core with no inertia or momentum within it.

Secondly, if the pressure in the core cannot increase beyond a certain level, then nature precludes the formational collapse leading to the development of the theoretically-proposed relativity-based Schwarzschild singularity. The U_{max} black shell star is the ultimate nemesis of Schwarzschild's singularity concept. In the light of absorption theory, the Schwarzschild singularity becomes redundant, unnecessary, and irrelevant.

The third fascinating new concept is to consider that when a star becomes a Stage 2 U_{max} star, what happens to the axial spin in its core? Since inertia is a function of acceleration, within a gravitational field and since spin is a form of acceleration, then, the Stage 2 core represents a unique environment in which there are no gravitational waves at all (this is fundamentally different from the zero net gravitational vector at the centre of a normal planet or star) and in which, therefore, there will be neither inertia nor momentum.

This is a unique and new concept to consider. The super-dense core material can now spin ultra-fast and can start and stop instantly. The core's speed of rotation can increase to a phenomenal value. This might be responsible for a jet of matter being expelled at the spin poles because of its freedom to react to trapped electromagnetic forces within the black shell. Electromagnetic oscillations in the core could give rise to the creation of pulsar properties. All of this speculation is included here because the author is the one person best equipped to make it. The author has been thinking about this concept for forty years and believes that the DOPA concept opens up the potential for centuries of research and development work.

132

5.6 The work on the possible evolution and construction of black shell stars (black holes) has highlighted the proposal that gravitational waves are the only wave type able to enter and then leave a black shell star through its black shell structure (event horizon).

This theory leads to a new concept and a new class of star altogether—the "*gravitational* shell star", which exerts the maximum possible gravitational force—a fixed U_{max}. Standard black shell stars are defined by their ability to retain EMF radiation; gravitational shell stars are defined by their ability to 'retain' gravitational waves through absorption.

The U_{max} concept may well, in the future, allow all other gravitational accelerations to be quoted as a percentage of Umax for any particular sector of the universe. i.e., A certain planet might have (as a fictitious example) a surface gravitational acceleration of 0.0001 U_{max} calibrated for the central sixty per cent volume of the known universe.

The theory that gravitational waves create the gravity force that holds together any cosmic body leads us to conclude that gravitational waves cannot be controlled by the very force that they, themselves, produce. Therefore, gravitational waves are not influenced by gravitational force and can thus not only pass into a standard black shell star in order to create its gravitational field but can pass out on the other side, leaving the star through the light (optical) event horizon that the gravitational force has generated. This may, in the far distant future, allow scientists to study gravitational waves emerging from black shell stars (black holes) to acquire information on the internal structures and activities within those stars' shells.

5.7 The author has investigated some of the weaknesses and failings in the mathematical equation that Newton's theory subsequently created to support his proposal for the gravitational force of attraction.

This work was not a direct consequence of Roberts' DOPA theory but is an interesting study that arose as a result of considering Newton's failure to provide a reasoned and explained mechanism for his proposal that the nature of gravitational force is an attraction between particles.

The outcomes described below do not cover the full range of work undertaken, or that could be undertaken. Primarily, the work described below covers an adjustment to the numerator of Newton's equation. More work is needed on the denominator. It is possible that, ultimately, such work could be incorporated into a later edition of this book.

Firstly, Newton's 'law' of gravitation was never written as a 'law', but was drawn from various of his statements which, when assembled, may be paraphrased to say:

Every particle in the universe attracts every other particle
with a force that is directly proportional to the product
of their masses, and inversely proportional to the square
of the distance between them.
—ISAAC NEWTON'S UNIVERSAL LAW OF GRAVITATION

133

However, what is not widely known is that **Newton never actually came up with the following equation**. It was attributed to him but was never written down in any of his works.

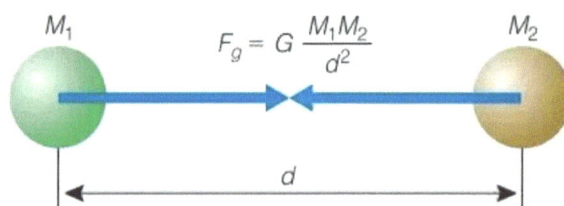

The author has not researched the history of the philosophy of gravity because it has already been researched more than amply by others. He prefers to quote from a book published in 2012 and written by Brian Clegg. It is entitled, "Gravity. Why What Goes Up Must Come Down." The following paragraphs, written in italics, are extracted from *Clegg, Brian. "Gravity" (Kindle Locations 853-893). Gerald Duckworth & Co. Kindle Edition.*

Amongst many other researched historical observations, Clegg wrote,

"...at the heart of Newton's third book is a formula that, though never explicitly used, can be deduced from Newton's various calculations and diagrams. It is now known as Newton's law of universal gravitation, and is usually presented in the form $F = Gm_1m_2 / r^2$.

"Newton didn't have a value for G** and only ever presented his law of gravitation by comparing two forces and how they varied. **In fact, it was not until 1894 that the British physicist Charles Vernon Boys came up with the concept of the universal gravitational constant, G, to provide the final version of the formula."

Clegg, Brian. Gravity (Kindle Locations 793-810). Gerald Duckworth & Co. Kindle Edition.

The equation, ostensibly, allows one to calculate the force between two particles, objects, planets, or stars. The units are not apparent, and there is a constant, G, whose units are abstruse and the equation's units, therefore, need defining carefully.

Any alteration of or change in the units given below will result in a nonsense output because the constant G only works with them in particular.

F is the force exerted between two bodies no matter how big or small.
Its units are Newtons. N. (Note NOT kN)

M_1 and M_2 are the masses of each of the two bodies respectively.
The units are kilogrammes. kg. *(Note NOT grammes or tonnes.)*

d is the distance between the centres of the two bodies.
The units are metres. m. *(Note NOT kilometres.)*

The most difficult one that usually prevents people from using this equation is Newton's gravitational constant G that has a complex unit whose construction is abstruse: $m^3.kg^{-1}.s^{-2}$ (metres to the power of 3, kg to the power of -1, and seconds to the power of -2.)

Fortunately, the units for G are of no real interest to the user because G is a constant and is therefore simply to be inserted into the equation as a number.

The number for G to be inserted into the equation is actually, $6.674215.E^{-11}$ according to *Jens H. Gundlach and Stephen M. Merkowitz 2000. Phys. Rev. Lett. 85, 2869.*

That is a very, very small number, being 6.674215 times 10 to the minus 11. But you cannot approximate it, because if you do, the equation does not work correctly. So, you literally insert it as stated here.

In using this equation, do not be tempted to make the units different from those specified. It will not work. At first, when using the equation, one thinks how absurd it is to have to insert the distance between, say, the Earth and the Moon in metres. But the distance must be inserted in metres even if you approximate the distance. For example, you do not need to specify the <u>exact</u> distance in metres as long as you put in a number in metres. And always remember that the distance between the objects is between the centres of spherical objects and not their outer surfaces.

For a test calculation, calculate the force on an 80 kg person standing on the surface of the Earth.
The mass of the Earth M_1 is $5.97.E10^{24}$ kg.   5.97 times ten to the 24th power.
The mass of the person M_2 is 80 kg
The distance between the two bodies is the radius of the Earth, which is usually taken as between 6,378,000 to 6,410,000 metres or thereabouts.
And the output force F that you should get should be about 9.81 Newtons. If it is different, say 9.70, then this is caused by the assumed radius and mass of the Earth. (Figures given in Wikipedia and other places vary for these two variables.)

Some history and philosophy related to gravity.

Whilst faithfully extracting Clegg's words" from his book, "Gravity", the author—for the benefit of the reader—has interjected his own commentary in non-italic square brackets.

"Remote influence is something that human beings inherently react against. When we see a magic trick where the magician appears to move something remotely, we assume there are wires or some other invisible but solid connection between the two.

In the final "general scholium" of Principia, Newton tells us that the motion of the heavenly bodies (and the tides) are down to gravity, **but that he has not decided on a cause for this gravitational pull.** [This is confirmation and support of the author's fundamental criticism of both Newton and Einstein's theories; that they could not find any mechanism for how their theories of motion operated—attraction in Newton's case and the behaviour of spacetime in Einstein's case.]

Because it acts in proportion to the total quantity of matter rather than the size of an object's surface, Newton deduces that this "gravity" thing can't be mechanical in the conventional sense. **However, he does not have a solution** *and— in the original Latin of Principia— he comments* ***'Hypotheses non fingo,'*** *which is traditionally translated as 'I frame no hypotheses.'*

Newton wasn't saying that he never uses hypotheses— in fact, as we have seen, book 3 of Principia starts with a bunch of hypotheses— but rather that **he is not prepared to make up some hypothesis for the cause of gravity. That word "fingo" is important as it doesn't mean something neutral like "propose." The Latin word (like the English "frame") is derogatory, suggesting making something up, inventing fictions, rather than conceiving in a thoughtful, scientific fashion.**

Newton concludes by saying that it is enough that gravity really exists and acts according to the laws he has described. In effect, as long as the math works to predict what happens in the world it doesn't really matter whether or not we understand why that outcome occurs. [This was the same conceit as exercised by Einstein. Neither could figure out the mechanism for their theories of how things behaved physically; but only Newton was bold enough to openly admit it. Einstein simply stated his view of how spacetime might function and left it to his acolytes to promote it.]

The whole thrust of scientific thinking in Newton's era, typified by Descartes' work, had been that everything was now explicable through mechanical means.

Newton made his feelings clear in a letter to classical scholar and theologian Richard Bentley. Written soon after the Principia was published, the letter declared that no one in their right mind could believe it was possible for a body to "act where it was not." His use of the term attraction was a way of describing the result, not a suggestion for the cause." [And this is where the author agrees with Newton and defies both him and Einstein. The reader will have realised that the practical effect of the Roberts theory is that matter does not respond to a force exerted between two objects or to a distortion in spacetime, but it responds to a pre-formed potential field created by the current passage of gravitational waves through space and matter. The ultimately important point being that the author has developed a theory for the thing that neither Newton nor Einstein could: the working mechanism by which gravitational force is created. Both Newton's and Einstein's theories of *matter behaviour can be based adequately on DOPA theory.]*

"Newton didn't have a value for G and only ever presented his law of gravitation by comparing two forces and how they varied. In fact, it was not until 1894 that the British physicist Charles Vernon-Boys came up with the concept of the universal gravitational constant, G, to provide the final version of the formula."
"Although **Newton never came up with the formula** *involving G, his work on gravitation was to have a huge impact— yet there was an issue that even he would acknowledge. He had no clue as to how gravity did its stuff."*

Clegg, Brian. Gravity (Kindle Locations 804-835). Gerald Duckworth & Co. Kindle Edition.

[Recognising—based on the above—that Newton never actually came up with the equation that is attributed to him as "Newton's Universal Law of Gravitation", for want of any better term, the author refers to it as "Newton's equation". Even though he never wrote it or used it. An amazing thought.]

Having looked at the philosophical and historical aspects of the concept of gravity, we can look at Newton's equation in more detail, which is the purpose of this section of the author's book.

It has to be wondered how Vernon-Boys developed Newton's equation and how he selected the components of his equation. After all, there are only four—which seems very few for such a significant equation that worked extremely well while it lasted. There is the mass of one object, the mass of a second, the distance between the centres of those two objects, and a constant to make the result come out the way that things actually work. That is all.

As far as developing his principles, it is clear that Newton was able to draw on the exploratory philosophies of many great scientists before him, but still it has to be recognised that for him to achieve his work on gravitation without the use of any modern calculating equipment such as computers, spreadsheets, or programming languages, without the astronomical information available today, and without any kind of modern infinitesimal calculus with which to undertake calculations, was a miraculous feat of genius. Nothing can detract from his immense achievements.

And yet, aspects of his work were questioned and, eventually rejected two hundred years later by the great genius Albert Einstein. Recognising the fallibility of Newton's work allows us to examine 'his equation' with objectivity. Perhaps it is truly because Newton himself didn't write the equation, only its principles, that some of its limitations and defects that are itemised by the author in this book were not publicised or listed previously. Therefore, the author considers that itemising such is a proper scientific service that is needed.

As a result of his examination, and as stated below, the author considers it is fair to say that some of the limitations inherent in the equation are, in fact, defects. Others are assumptions that can be questioned.

The biggest problem with Newton's law relates to the lack of a working mechanism. Had Newton actually written down his law as an equation, then he might have recognised the defects in it. But he did not.

Addressed below, are some points of criticism that need to be recognised and considered.

Point 1 needing consideration: Action across the universe.

Let us start with the title that has been attributed to it. "Isaac Newton's Universal Law of Gravitation". The word 'universal' in this context does not mean that it 'does everything'. Rather, it means that it applies right across the universe. Which it does attempt to do.

The problem with this proposal is that the calculations become trapped in the immense distances involved and the need for actions and reactions to take place over such great distances in a meaningful way. After all, the law states that every single particle in the universe attracts every other particle—but with what time lag? If we accept that the speed of light is the fastest speed that anything can travel in a vacuum, then we are looking at hundreds of billions of light years' time lag between, for example, one atom on a distant galaxy recognising that an atom on earth has been destroyed in a nuclear explosion.

For Newton's equation to work universally, as in its title, its effect must work faster than the speed of light - instantaneously, in fact. Impossible, unfortunately.

This is the first problem with Newton's equation. It doesn't work as advertised! Which is a pretty big problem to exist in an equation that has been used and accepted for so long.

So, what does this failing highlight? It highlights the fact that Newton's law of gravitation, as expressed in his equation, simply describes how things move. It does not contain any explanation for HOW it works. There is no mechanism inherent in its construction.

That is exactly the same failing in Einstein's spacetime concept. It relies on relativity to describe how things move, but there is no working mechanism to underpin HOW it works.

Point 2 needing consideration: The use of an inverse d^2 term.

Why did Newton decide to use such a simple term as an inverse d^2 to describe the effect of distance between objects to provide a prime variation in the force between those objects?

Newton would be aware, as indeed must the reader, that the inverse d^2 term extends the equation to infinity. He had to have realised that this made his equation 'universal' and thus impossible to verify and impossible to justify.

Newton's general explanation is to use the d^2 term to reflect the spherical nature of dissipation of his attraction force. It is a standard enough concept based on the area of the sphere as the sphere of influence expands. However, it is unacceptable as a 'cover story', because Newton specifically postulated as a critical part of his theory that each atom is 'attached' to every other atom in the universe by a mysterious attractive force, not that there is a general field that extends with decreasing intensity to infinity interacting as a general field with every atom that it encounters on the way.

The expanding sphere concept does not match up with Newton's individual atom-to-atom, connected-attraction, proposal. He did not, and could not, justify why the attraction between two atoms should reduce at the inverse rate of d^2. It seems to the author that, if one has a linear connection between any two given particles, then the force between the two might weaken linearly in direct inverse proportion to the distance between the particles, not to the square power.

It strikes the author that a reduction in a direct connecting force between two objects should more realistically reduce at an inverse linear rate. If the sphere concept has to be invoked, which the d^2 does, then it is not appropriate to describe the attraction effect as existing between individual particles. It invalidates the equation.

The use of this expression may well produce a realistic and verifiable result with relatively short distances such as those within the inner part of our solar system, but longer-distance results have still been impossible to check, and, after all, the fact that the relativity mathematics has produced more accurate results is indicative that there must be structural errors in the 'universal' equation.

The author has not yet experimented with altering this variable because he has found that his adjustment of the numerator mass variables in the equation (see below) can bring down the closing of the equation significantly, coming in from infinity, thus making the author's new equation's outcomes potentially more realistic. These are described below, but that does not mean that the equation cannot be improved by an improved distance variable expression. It probably can. The author's expected outcome is that the universal equation is likely to be capable of being re-written significantly to provide an improved outcome performance.

[The author considers that what Newton was struggling to achieve (as indeed was Einstein) was to find a way to create an infinitely-extending gravitational field that somehow caused objects **to form and move together in response to their individual masses**. Both of them failed to achieve their objective. The author takes this opportunity to point out that the Roberts potential gravitational field developed in DOPA theory provides that which both Newton and Einstein sought—a response field that allows things to create and respond to gravitational force without experiencing the problems inherent in 'action at a distance'. DOPA theory creates the gravity zones described by Newton and described by Einstein, but for which they could provide no mechanism. Absorption theory provides a pre-existing potential gravitational field caused by the imbalance in the gravitational wave flux, which would control the behaviour of any mass inserted into it. This, effectively replaces Newton's universal law of gravitational attraction and contributes a foundation to Einstein's spacetime by providing the long-sought working mechanism.]

Point 3 needing consideration: The use, in the numerator, of the product of the masses of the objects under consideration.

This is the major query as to the structural soundness of Newton's 'universal' principles. Why did Newton consider that the product of the individual masses of the two objects under consideration was the best expression to obtain the correct force between bodies? Why, indeed, did he even consider it to be a valid expression?

The author considers, firstly, that it seems most odd and unlikely, that an outcome dependent on the mass of one object in relation to the mass of another object should relate to their product. Would it not seem much more logical that, somehow, if gravitational attraction was related to their masses, then it should, logically, be somehow related to the sum of their masses?

The first problem related to this expression of $M_1 \times M_2$ is that it, effectively, prevents the insertion of the values in a meaningful way. By this, the author means that the product of $M_1 \times M_2$ can be derived by several different multipliers, which means that the result of the multiplication, which should be unique, is not so. In this case, M1 x M2 is meaningless.

Consider the incorrect saying that is commonly used:

Technically - gravity is an attraction between TWO masses (one of which is the Earth). A small feather pulls on the Earth with exactly the same force that the Earth pulls on it...of course since force is mass times acceleration - the acceleration of the Earth towards the feather is microscopic - where the acceleration of the feather is $9.8 m/s^2$ ...but the FORCE exerted by the feather is the same as the earth exerts on the feather.

Surely, if the alleged 'pull' between the Earth and a feather is identical and is controlled by the mass of the feather, then how can the product of their masses be relevant at all?

And, Newton does not say why the two should pull on each other with the same force. He does not ascribe any mechanism to that rule.

To study this matter, we need to start at the very smallest scale and consider what Newton did say when he described his theory of attraction. He said that every particle (atom) in the universe attracts every other particle with a force that is directly proportional to the product of their masses, and inversely proportional to the square of the distance between them.

(In saying the above, he did not mention an absolute value, only proportional values because he had not developed and found a constant that would complete the equation. Also, he did not mention that d is the distance between the centres of spherical bodies such as planets or stars; why, the author does not know.)

When we discuss the equation against Newton's theory, we are discussing two things separated by time, but one of which—the equation—has come to represent Newton's theory and quantify it. The problem is to study whether the theory is sound, as promoted by Newton, and whether the equation that has evolved from the theory is sound, as used today.

So, we must start at the smallest scale to consider the validity of what Newton said.

Newton's theory atom by atom

What if we suggest, in order to help him, that Newton's expressed law really meant that each atom can only generate a given attraction of its own no matter how many other atoms it tries to attract? In that case, we are working on a different basis. That would sound sensible. After all, in our everyday experience, a magnet can only generate a certain maximum pulling power no matter how big or small an object it tries to move.

So, what would happen in such a case?

We could propose that the maximum Newtonian attraction force an atom can exert is 1 unit of force, whatever that value might be, as described above.

So, let us say that this atom (which we can label A) tries to attract an adjacent body comprised of 11 atoms identical to itself at a distance of 1 metre (atoms labelled B to K). Its single maximum unit of force must, surely, be distributed equally between all of the 11 atoms of the other body. It cannot conjure up 11 units of force, one for each atom. That is reasonable otherwise it would have to generate a near-infinite amount of force in its attraction process with all the other atoms in the universe. Thus, each atom of the larger, adjacent body experiences $1/11^{th}$ of a maximum force unit and exerts a $1/11^{th}$ force back on atom A. In this case, the general principle that Newton proposed—each body attracts each other body with an equal force—would apply. Both 'bodies' would experience a single maximum attraction force unit.

That all seems satisfactory until we examine the picture more closely.

The pulling power of each of atoms B to K is now reduced by $1/11^{th}$ from each atom, and, irrespective of nuclear strong forces and similar, according to Newton, each atom must be attracting each of the other 11 atoms to itself, but with a considerably increased force because they are microscopically close (bear in mind the 'inverse square of the distance' part of the principle of attraction). This starts to create a conundrum. And the conundrum gets worse when we reflect on the proposition made by Newton that every atom is in direct contact and holds an attractive connection to every other single atom in the universe! Does the inverse square law mean that the total force being exerted by atom A is infinity or 1? Or 2? Or what? The reader can see that this rapidly becomes nonsensical.

We need to see whether an unlikely alternative will make more sense. What if an atom can only exert 1 unit of force amongst all the other atoms of the universe, but can receive all the force from all the other atoms whether far or near? No, that doesn't make sense at all. When you start to examine it, whichever way you look at it, Newton's proposal is not at all acceptable.

But, let's struggle on and just consider the problem of the 12 atoms and ignore the highly confusing matter of attraction force between particles and move on to consider each particle just as having mass. That is, essentially, what Newton's equation does. It says that force between atoms depends on the mass of any two bodies being considered. If we ignore the universal constant, which we can do in comparative statements, and we assume a distance of 1 metre between the bodies to remove the matter of inverse d squared, we recognise Newton's equation as $F = m1 \times m2$.

Newton said that the force between any two particles or bodies is derived directly from their masses. If their masses are multiplied, we have the outcome force. That's what he said.

With m1 being 11 atoms and m2 being 1 identical atom, we have $F = 11 \times 1 = 11$ units of 'mass related' force. 12 atoms attracting each other with a force of 11 units of mass.

If, instead, we take two bodies comprising of 10 atoms and 2 atoms, then we have $F = 10 \times 2 = 20$ units of force. The same 12 atoms attracting each other with a force of 20 units. That makes no sense.

If, instead, we take two bodies comprising of 9 atoms and 3 atoms, then we have $F = 9 \times 3 = 27$ units of force. The same 12 atoms attracting each other with a force of 27 units.

And $8 \times 4 = 32$ units, and $7 \times 5 = 35$ units, and $6 \times 6 = 36$ units.

Does that make any sense at all? Not really. And yet this is what Newton claimed. That the force is obtained by multiplying the mass of the individual units in kg. Not any other units but kg. And that is directly related to the number of atoms contained in the body, assuming that the atoms are the same for the purposes of the discussion.

This detailed examination of the principles of Newton's theory leads us to our first glimpse of the possibility that his equation may be impossible and faulty.

How can we start to show that? If the reader considers that talking about forces and how they do not work, then the only way to make the point is to demonstrate it by moving over to actual mass numbers using Newton's actual full equation and taking examples that give us a different perspective. In the examples using atoms, we kept the total mass the same and varied the number of constituent atoms. Below, we vary the product mass of the objects and show how a wide range of objects can produce the same output force. That would be a nonsensical result, of course.

Consider various bodies conforming to $F = G.M_1.M_2/d^2$

$M_1 = 3.162$ kg. $M_2 = 3.162$ kg. $d = 1$ metre $F = 10G$ Newtons
Total mass 6.324 kg.

$M_1 = 10$ kg. $M_2 = 1$ kg. $d = 1$ metre $F = 10G$ Newtons
Total mass 11 kg.

$M_1 = 10,000$ kg. $M_2 = 0.001$ kg. $d = 1$ metre $F = 10G$ Newtons
Total mass 10,000.001 kg.

It is clear that the above three examples highlight the inconsistency of the equation. It also precludes any back-analysis from an output force value. It highlights the flaw inherent in the quotation where the obvious question is, **"Under Newton's equation, how do the Earth and the feather know how much force to exert on each other when the force is the product of their masses and not related to their individual masses?"**

What this defect says it that each of these three disparate 'body pairs' above attract each other with the same force. A 10,000 kg object exerts the same 10G Newton force on a 0.001 kg 'feather' as does a 10 kg object on a 1 kg 'feather', or two equal-masses weights of 3.162 kg. Thus, a total mass of 6.324 kg. is supposed to exert a mutually equal force to a total mass of 10,000.001 kg. How does that seem at all reasonable?

And so, what about absorption theory and the new equation?

Absorption theory explains above why bodies of different size can and do attract each other with equal force (Fig 35), but the mathematics behind Newton's presentation is flawed in that the amount of the force is not consistently obtained and is not controlled by the relative masses of the objects. The Newtonian theory does not provide any proper explanation for why it should be so. Absorption theory does.

Absorption theory succeeds because it separates the two bodies from each other, each one creating its own internal drag towards the other based on each one's own mass. Yet the force being

142

experienced by the pair is (as illustrated in Figure 35) controlled by the smaller, less massive, of the two objects.

In order to see whether this situation could be improved, the author made some preliminary studies and quite quickly realised that the numerator in the universal equation could be adjusted to provide at least an ability to adjust the values of M_1 and M_2 separately to consider varying outcomes. The numerator became $M_2(M_1+M_2)$ where M_2 had to be the smaller mass of the two. The author's new equation separates the two bodies and identifies them so that the outcome of calculations may be followed and determined properly. For the first time, we are now able to at least obtain a qualitative concept of what will happen when M_1 and M_2 vary such that, for example, they become nearly the same mass. Are there any fundamental differences produced by their ratios? The author has investigated and describes below some initial outcomes of interest.

Point 4 needing consideration: The equation does not cater for the radii of the two objects, thus precluding any inbuilt limitation on their proximity.

This is a simple matter that there are many geometries where the two objects could be touching and the equation does not cater for this. It is clear that this is a defect that could easily be catered for by defining mass as volume and density and hence catering for their mutual radii (see Point 6).

Point 5 needing consideration: The equation does not cater for the limiting of acceptable radii of proximity in order to avoid two objects entering into each other's Roche limit.

There is a distance proximity limit within which massive objects will detrimentally affect each other through their mutual gravitational force. This is called the Roche limit (See Section 3.8). Considering that Newton's equation is an equation defining gravitational force, it seems highly inadequate that it does not cater for this important restriction which controls whether or not a minor body will be ripped apart by a larger body if they become too close to each other. There is also the problem of the total mass of the bodies because there is a total mass limit within which Roche forces do not damage objects. For example, there is no Roche limit on two 1 kg. objects approaching each other. This harks back to the inadequacy of an equation that does not specify mass as a function of size, volume, and density (see Point 6).

Point 6 needing consideration: The equation does not permit any analysis of the contributions of the individual masses of the objects to their mutual gravitational attraction.
There was, clearly, a need to change the equation so that one could alter the values of the mass of each of the two bodies and look at the outcome differences. So, the author created an alternative equation and, as a starting point, developed the following modified equation:

$$F = G_R.M_L.(M_G+M_L)/d^2$$

Where d is the centre-to-centre distance in metres
M_G is the object with the Greater mass (measured in kg)
M_L is the object with the Lesser mass (measured in kg)
$d > R_G+R_L$
$d >$ than the Roche limit for the lesser mass body M_L approaching the greater mass M_G

G_R is Roberts' gravitational constant where the Newtonian value of G can be changed to G_R for research purposes. In that regard, G_R might also be now considered as a variable.

This new equation has allowed the author to examine results based on variations in the actual and relative sizes of the two objects. Newton's equation did not allow this. And yet, this simple alteration produces the same results as Newton's equation for most practical purposes.

The author's equation now also permits the adjustment of the mass variables to adjust and control their masses through specification of their radii and densities using the substitution of $4/3pi.R^3$ for M_G and M_L. This facility will, then, also permit the user to control the limit on the variable 'd' in relation to R_S+R_L in order to ensure that the objects neither touch one another nor enter into their Roche zone. The author is not aware of the minimum mass needed for an object to be susceptible to gravitational disruption, but other researchers may be aware of this value.

The author proposes that the development of this new equation calculating the gravitational force between two bodies is a significant step towards a better understanding of the behaviour of matter in response to gravitational force and that this will lead to new ways of investigating and forecasting the effects of gravity.

Using nothing more complicated than a spreadsheet, we can see some interesting results obtained by comparing Newton's and Roberts' equations.

VARIABLE ONE-OFF SCENARIO COMPARISON BETWEEN NEWTON'S AND ROBERTS' OUTPUT
M_G = GREATER MASS OBJECT   M_L = LESSER MASS OBJECT   d = DISTANCE APART CENTRE TO CENTRE
STUDY COMPARISON OF EQUATION OUTPUT CONSEQUENT UPON EQUATION DIFFERENCES

HAND INPUT DATA COLOURED RED   HAND INPUT DATA COLOURED RED

$F = G.M_G.M_L/d^2$ (Newton's)	$F = G_R.M_L.(M_G+M_L)/d^2$ (Roberts')
TEST DATA INPUT	**TEST DATA INPUT**
6.67E-11 G Constant	6.67E-11 G Constant
1.99E+30 MG kg Sun	1.99E+30 MG kg Sun
7.35E+22 ML kg Moon	7.35E+22 ML kg Moon
1.50E+11 d centre to centre metres	1.50E+11 d centre to centre metres
4.3562410E+20 Force F in Newtons	4.3562412E+20 Force F in Newtons
4.36E+17 F in kN	4.36E+17 F in kN
4.44E+19 F as kg	4.44E+19 F as kg
0.00593 acceleration of gravity m/sec2	0.00593 acceleration of gravity m/sec2
ON EARTH	**ON EARTH**
1 N = = 0.101972 kg	1 N = = 0.101972 kg
1kN = = 101.97 kg	1kN = = 101.97 kg
1 kg = = 9.81 N	1 kg = = 9.81 N

Table 5.7.1 Both Newton and Roberts' equations produce the same result for the force of gravitation between the Sun and our Moon. 4.36 E^{+20} Newtons identical to a difference of 1:20 million.

Consider Table 5.7.1, where the relevant inputs have been made to calculate the force between the Sun and the Moon. It can be seen that both Newton's and the author's equations produce the same output of 4.36 E^{+20} newtons when taken to two decimal places. However, the results are demonstrably the same to six decimal places, differing slightly in only the seventh decimal place. Newton gives 4.3562410 E^{+20} newtons while the author gives 4.3562412 E^{+20} newtons. That is the results are identical to better than one part in twenty million. And—this is most important to notice—there is no evidence available to say that Newton's equation is the one that produces the most accurate output. It is equally reasonable to propose that the author's equation produces the correct output and that Newton's is the equation that is incorrect to one twenty millionth part. Especially considering the simpler construction of Newton's equation.

This seems to be a remarkable achievement, particularly when one considers that, to the best of the author's knowledge, no alternative to Newton's equation has been proposed over the last three hundred years and bearing in mind that this analysis has been performed specifying the masses of the two bodies in such a way as to make the analysis unique and not confusable. Of course, it is wonderful to recognise that this new equation opens up the result to further comparative analysis.

For example, as mentioned above, the values of mass can be further broken down to elaborate the author's equation by allowing the direct input of a planet's radius and average density. Thus:

$$F = G_R.M_L.(M_G+M_L)/d^2$$

can become

$$F = G_R.[M]_L.([M]_G+[M]_L)/d^2$$

If we allow M to be replaced by the volume multiplied by the density of each object, then we shall obtain the mass of each object/planet by specifying two new variables: R, the radius of the object/planet in metres and ρ the average density of the object/planet in g/cm3 or kg/m3.

So: R_G is the radius in metres of the object with the greater mass
And ρ_G is the average density of the object with the greater mass

Note that, because of the potential range of the variables, it is possible that R_G could be smaller than R_L if ρ_G were sufficiently large that their product was a greater mass than that of the lesser massive object. The calculation is somewhat like an iterative solution in that you will not know which object is the more massive of the two until you have performed the two sets of calculations. First calculate R. ρ for each of the two objects, and whichever is the more massive acquires the G suffix. Whichever is the least massive acquires the L suffix. That is straightforward but has been spelt out here for the avoidance of any doubt.

Where G_R is the Roberts gravitational constant which can be the same as Newton's G or, if required for analytical purposes, can be altered to suit. In the calculations illustrated in this book, $G_R = G$

Where 4/3 Pi = 4.18879 and ρ is the average density of the object/planet
And 4/3pi.R_G^3 is the volume of the greater mass object/planet
And 4/3pi.$R_G^3.\rho_G$ is the mass of the greater mass object/planet
And the average density of each body is denoted by ρ_G and ρ_L

Thus:

$F = G_R.[M]_L.([M]_G+[M]_L)/d^2$ can become:

$F = G_R. [4.18879.R_L{}^3.\boldsymbol{\rho}_L] . ([4.18879.R_G{}^3.\boldsymbol{\rho}_G] + [4.18879.R_L{}^3.\boldsymbol{\rho}_L])/d^2$

$F = G_R.17.5459 [R_L{}^3.\boldsymbol{\rho}_L] . ([R_G{}^3.\boldsymbol{\rho}_G] + [R_L{}^3.\boldsymbol{\rho}_L])/d^2$

Thus, Roberts' gravitational equation may be expressed as either

$F = G_R.M_L.(M_G+M_L)/d^2$

or

$F = G_R.17.5459 [R_L{}^3.\boldsymbol{\rho}_L] . ([R_G{}^3.\boldsymbol{\rho}_G] + [R_L{}^3.\boldsymbol{\rho}_L])/d^2$

depending upon whether you wish to use just the mass of each of the two objects directly, or whether you would prefer to calculate using the volume and density of the two objects to produce their mass.

And, it goes without saying that one can use any mixture of the two notations to arrive at the correct answer.

In the spreadsheet illustrations included in this book, in addition to the use of the Newton equation, comparative calculations have been undertaken using the author's expression $F = G_R.M_L.(M_G+M_L)/d^2$ in order to explore a different format from Newton's equation. In Roberts' equation, G_R is a gravitational constant that may be selected at different values for experimentation purposes; the R denoting Roberts. M_L is the mass of the lesser massive body and M_G is that of the greater massive body, both in kg. d is the distance between the centres of the bodies in metres.

OBSERVATIONS

This scenario examines the effect on the force generated between two objects (one of which is the Earth (MG) and the other a 1,000 tonne satellite ML), as the distance (d) is increased between them. At a separation of 835 billion km, the Roberts equation holds true to 0.001 Newtons. The upper set of calculations are based on Newton's equation, and the lower set is based on Roberts' equation. The inverse d-squared denominator controls the outcome irrrespective of the numerator composition. It is interesting to note that 1 newton is about the weight of a typical apple. So the gravimetric force on the thousand-tonne satellite is tiny indeed but still felt. Also, to add some context, the distance of 835 million km is a little bit more than the distance of Jupiter from the Sun. The first 'd' reading on the list is the distance is the Earth's radius to its surface, which validates the calculations bu virtue of the calculated 9.80 m/sec2 acceleration.

Newton's 2nd law. f = m.a, where 'a', for example, on the surface of the Earth IS 9.80665 m/sec2

CALCULATIONS USING NEWTON'S EQUATION $F = G \times MG \times ML / d^2$

	G constant N(m/kg)2	MG kg	ML kg	d metres	Force F Newtons	grav. accel (m/sec2)	Ratio of ML/MG as %
0	6.674E-11	5.97E+24	1,000,000	6,378,000	9,797,989.542	9.80	0.000
1	6.674E-11	5.97E+24	1,000,000	12,756,000	2,449,497.385	2.45	0.000
2	6.674E-11	5.97E+24	1,000,000	25,512,000	612,374.346	0.61	0.000
3	6.674E-11	5.97E+24	1,000,000	51,024,000	153,093.587	0.15	0.000
4	6.674E-11	5.97E+24	1,000,000	102,048,000	38,273.397	0.04	0.000
5	6.674E-11	5.97E+24	1,000,000	204,096,000	9,568.349	0.01	0.000
6	6.674E-11	5.97E+24	1,000,000	408,192,000	2,392.087	0.00	0.000
7	6.674E-11	5.97E+24	1,000,000	816,384,000	598.022	0.00	0.000
8	6.674E-11	5.97E+24	1,000,000	1,632,768,000	149.505	0.00	0.000
9	6.674E-11	5.97E+24	1,000,000	3,265,536,000	37.376	0.00	0.000
10	6.674E-11	5.97E+24	1,000,000	6,531,072,000	9.344	0.00	0.000
11	6.674E-11	5.97E+24	1,000,000	13,062,144,000	2.336	0.00	0.000
12	6.674E-11	5.97E+24	1,000,000	26,124,288,000	0.584	0.00	0.000
13	6.674E-11	5.97E+24	1,000,000	52,248,576,000	0.146	0.00	0.000
14	6.674E-11	5.97E+24	1,000,000	104,497,152,000	0.037	0.00	0.000
15	6.674E-11'	5.97E+24	1,000,000	208,994,304,000	0.009	0.00	0.000
16	6.674E-11	5.97E+24	1,000,000	417,988,608,000	0.002	0.00	0.000
17	6.674E-11	5.97E+24	1,000,000	835,977,216,000	0.001	0.00	0.000
18	6.674E-11	5.97E+24	1,000,000	1,671,954,432,000	0.000	0.00	0.000
19	6.674E-11	5.97E+24	1,000,000	3,343,908,864,000	0.000	0.00	0.000
20	6.674E-11	5.97E+24	1,000,000	6,687,817,728,000	0.000	0.00	0.000

CALCULATIONS USING ROBERTS' EQUATION $F = GR \times ML \times (MG + ML) / d^2$

	GR constant N(m/kg)2	m1 kg	m2 kg	d metres	Force F Newtons	grav. accel (m/sec2)	Ratio of ML/MG as %	Ratio of force F for two equations Newton/Roberts
0	6.674E-11	5.97E+24	1,000,000	6,378,000	9,797,989.542	9.80	0.000	1.000000000000
1	6.674E-11	5.97E+24	1,000,000	12,756,000	2,449,497.385	2.45	0.000	1.000000000000
2	6.674E-11	5.97E+24	1,000,000	25,512,000	612,374.346	0.61	0.000	1.000000000000
3	6.674E-11	5.97E+24	1,000,000	51,024,000	153,093.587	0.15	0.000	1.000000000000
4	6.674E-11	5.97E+24	1,000,000	102,048,000	38,273.397	0.04	0.000	1.000000000000
5	6.674E-11	5.97E+24	1,000,000	204,096,000	9,568.349	0.01	0.000	1.000000000000
6	6.674E-11	5.97E+24	1,000,000	408,192,000	2,392.087	0.00	0.000	1.000000000000
7	6.674E-11	5.97E+24	1,000,000	816,384,000	598.022	0.00	0.000	1.000000000000
8	6.674E-11	5.97E+24	1,000,000	1,632,768,000	149.505	0.00	0.000	1.000000000000
9	6.674E-11	5.97E+24	1,000,000	3,265,536,000	37.376	0.00	0.000	1.000000000000
10	6.674E-11	5.97E+24	1,000,000	6,531,072,000	9.344	0.00	0.000	1.000000000000
11	6.674E-11	5.97E+24	1,000,000	13,062,144,000	2.336	0.00	0.000	1.000000000000
12	6.674E-11	5.97E+24	1,000,000	26,124,288,000	0.584	0.00	0.000	1.000000000000
13	6.674E-11	5.97E+24	1,000,000	52,248,576,000	0.146	0.00	0.000	1.000000000000
14	6.674E-11	5.97E+24	1,000,000	104,497,152,000	0.037	0.00	0.000	1.000000000000
15	6.674E-11	5.97E+24	1,000,000	208,994,304,000	0.009	0.00	0.000	1.000000000000
16	6.674E-11	5.97E+24	1,000,000	417,988,608,000	0.002	0.00	0.000	1.000000000000
17	6.674E-11	5.97E+24	1,000,000	835,977,216,000	0.001	0.00	0.000	1.000000000000
18	6.674E-11	5.97E+24	1,000,000	1,671,954,432,000	0.000	0.00	0.000	1.000000000000
19	6.674E-11	5.97E+24	1,000,000	3,343,908,864,000	0.000	0.00	0.000	1.000000000000
20	6.674E-11	5.97E+24	1,000,000	6,687,817,728,000	0.000	0.00	0.000	1.000000000000

Table 5.7.2 A comparative calculation of the gravitational force between the Earth and a 1,000 tonne vehicle at increasing distances from the Earth. Agreement between Newton and Roberts is better than 0.001 Newtons force at 835 million km distance.

Table 5.7.2 shows a spreadsheet output showing the gravitational force generated between the Earth and a 1,000 tonne space vehicle at increasing distances from the Earth. The upper table shows the output results from calculating using Newton's equation and the lower set of figures shows the output using the author's equation.

There is no discernible or meaningful difference between the Newtonian and the Robertsian equations' outputs.

Just to demonstrate the validity of both equations, each of the two analytical runs starts off at the surface of the Earth so that the reader can see that the correct gravitational force is produced when d is the Earth's radius. It would be straightforward for the reader to produce his/her own equivalent spreadsheet inserting the relevant equations as required in order to conduct their own verifying analyses.

Perhaps the most important, and yet most obvious, observation to be made from the above simple spreadsheet table (which is available for everyone to experiment with) is that, by the time the million kg spaceship is somewhere out just beyond the orbit distance of Jupiter from the Sun, the Earth's gravitational force on it is less than one thousandth of the Earth weight of an apple! About the weight of an apple pip! In practical terms, it means that, for all practical purposes, the vehicle has escaped the pull of the Earth long ago as it travelled away.

And this means, again in practical terms, that it is nonsense to talk about the force of gravity being an attraction that extends out to infinity and still existing. Such is the statement of Newton in using a single universe-wide field concept of attraction to describe his theory about gravity.

Conversely, DOPA theory does not require a universe-wide single field. Instead, it relies on the rapidly-diminishing effects of the gravitational wave shadow effect between any two objects. Indeed, the table 5.7.2 supports DOPA theory by demonstrating firmly that the rapidly-diminishing effects of the gravitational shadowing based simply on the geometry of the situation are realistic and would work. It was just bad luck for Newton that he did not come up with the geometric shading concept at the time. But, then again, philosophical and religious perspectives were very much different from ours today; the occult was something much more real to them at that time, and Newton's invisible force of attraction did not seem so outlandish to him. And, since it appeared to work satisfactorily, as described by Brian Clegg in his book, people simply accepted that attraction must have been the way it all worked.

Fortunately, the author's simple theory based on the geometry of shading, as exemplified in the many diagrams contained in this book, needs no occult belief to function, and to function exactly as Newton forecast his attraction would. But there is no attraction. Which is particularly useful, because the shadow between the Earth and the thousand tonne spaceship is demonstrated by Table 5.7.2 to have become completely negligible long before the spaceship reached Jupiter's orbit. DOPA theory offers no nonsense about the effects of gravity extending to the infinite limits of the universe.

The more pragmatic conclusion that must be drawn and recognised from that table is that gravity is, in reality, a very localised mechanism on an interstellar scale. *It works, but only just*, one might say. The many small moons of Jupiter are so free of the Sun's influence that they are not affected by the Sun for any practical purpose—not when compared with the much more local effect of Jupiter's own gravitational shadow field between it and them. At that distance, the force that the Sun exerts is reduced to apple pips!

Roberts' new equation, $F = G_R.M_S.(M_L+M_S)/d^2$, works perfectly to describe the behaviour of gravitational force on matter throughout the Solar System and, beyond.

And there are yet more interesting spreadsheet tables to discuss, as may be examined below
.

OBSERVATIONS

From the Earth at a distance of 400,000 km, the Roberts equation holds to better than 17 decimal places, for all satellites up to 3,300 tonnes.
The distance of 400,000 km is just beyond the orbit of the Moon. This range has been selected to study the possibility that, one day,
we may wish to place satellites up to, say, 3,000 tonnes of mass into close-lunar orbits. Roberts' equation would place them accurately.

Newton's 2nd law. f = m.a, where 'a', for example, on the surface of the Earth IS approximately 9.81 m/sec2

CALCULATIONS USING NEWTON'S EQUATION   F = G x MG x ML / d^2

	G constant N(m/kg)2	MG kg	ML kg	d metres	Force F Newtons	grav. accel (m/sec2)	Ratio of ML/MG as %	
0	6.674E-11	5.97E+24	1,000	6,378,000	9,798	9.7980	0.000	Earth demonstration line
1	6.674E-11	5.97E+24	1,500	400,000,000	4	0.0025	0.000	
2	6.674E-11	5.97E+24	2,250	400,000,000	6	0.0025	0.000	
3	6.674E-11	5.97E+24	3,375	400,000,000	8	0.0025	0.000	
4	6.674E-11	5.97E+24	5,063	400,000,000	13	0.0025	0.000	
5	6.674E-11	5.97E+24	7,594	400,000,000	19	0.0025	0.000	
6	6.674E-11	5.97E+24	11,391	400,000,000	28	0.0025	0.000	
7	6.674E-11	5.97E+24	17,086	400,000,000	43	0.0025	0.000	
8	6.674E-11	5.97E+24	25,629	400,000,000	64	0.0025	0.000	
9	6.674E-11	5.97E+24	38,443	400,000,000	96	0.0025	0.000	
10	6.674E-11	5.97E+24	57,665	400,000,000	144	0.0025	0.000	
11	6.674E-11	5.97E+24	86,498	400,000,000	215	0.0025	0.000	
12	6.674E-11	5.97E+24	129,746	400,000,000	323	0.0025	0.000	
13	6.674E-11	5.97E+24	194,620	400,000,000	485	0.0025	0.000	
14	6.674E-11	5.97E+24	291,929	400,000,000	727	0.0025	0.000	
15	6.674E-11	5.97E+24	437,894	400,000,000	1,091	0.0025	0.000	
16	6.674E-11	5.97E+24	656,841	400,000,000	1,636	0.0025	0.000	
17	6.674E-11	5.97E+24	985,261	400,000,000	2,454	0.0025	0.000	
18	6.674E-11	5.97E+24	1,477,892	400,000,000	3,682	0.0025	0.000	
19	6.674E-11	5.97E+24	2,216,838	400,000,000	5,522	0.0025	0.000	
20	6.674E-11	5.97E+24	3,325,257	400,000,000	8,283.449	0.0025	0.000	

CALCULATIONS USING ROBERTS' EQUATION   F = G x ML x (MG + ML) / d^2

	GR constant N(m/kg)2	MG kg	ML kg	d metres	Force F Newtons	grav. accel (m/sec2)	Ratio of ML/MG as %	Ratio of force F for two equations Newton/Roberts	
0	6.674E-11	5.97E+24	1,000	6,378,000	9,797.99	9.7980	1.67E-20	1.000000000000	Earth demonstration line
1	6.674E-11	5.97E+24	1,500	400,000,000	3.74	0.0025	2.51E-20	1.000000000000	
2	6.674E-11	5.97E+24	2,250	400,000,000	5.60	0.0025	3.77E-20	1.000000000000	
3	6.674E-11	5.97E+24	3,375	400,000,000	8.41	0.0025	5.65E-20	1.000000000000	
4	6.674E-11	5.97E+24	5,063	400,000,000	12.61	0.0025	8.48E-20	1.000000000000	
5	6.674E-11	5.97E+24	7,594	400,000,000	18.92	0.0025	1.27E-19	1.000000000000	
6	6.674E-11	5.97E+24	11,391	400,000,000	28.37	0.0025	1.91E-19	1.000000000000	
7	6.674E-11	5.97E+24	17,086	400,000,000	42.56	0.0025	2.86E-19	1.000000000000	
8	6.674E-11	5.97E+24	25,629	400,000,000	63.84	0.0025	4.29E-19	1.000000000000	
9	6.674E-11	5.97E+24	38,443	400,000,000	95.77	0.0025	6.44E-19	1.000000000000	
10	6.674E-11	5.97E+24	57,665	400,000,000	144	0.0025	9.66E-19	1.000000000000	
11	6.674E-11	5.97E+24	86,498	400,000,000	215	0.0025	1.45E-18	1.000000000000	
12	6.674E-11	5.97E+24	129,746	400,000,000	323	0.0025	2.17E-18	1.000000000000	
13	6.674E-11	5.97E+24	194,620	400,000,000	485	0.0025	3.26E-18	1.000000000000	
14	6.674E-11	5.97E+24	291,929	400,000,000	727	0.0025	4.89E-18	1.000000000000	
15	6.674E-11	5.97E+24	437,894	400,000,000	1,091	0.0025	7.33E-18	1.000000000000	
16	6.674E-11	5.97E+24	656,841	400,000,000	1,636	0.0025	1.10E-17	1.000000000000	
17	6.674E-11	5.97E+24	985,261	400,000,000	2,454	0.0025	1.65E-17	1.000000000000	
18	6.674E-11	5.97E+24	1,477,892	400,000,000	3,682	0.0025	2.47E-17	1.000000000000	
19	6.674E-11	5.97E+24	2,216,838	400,000,000	5,522	0.0025	3.71E-17	1.000000000000	
20	6.674E-11	5.97E+24	3,325,257	400,000,000	8,283.449	0.0025	5.57E-17	1.000000000000	

Table 5.7.3 A comparative calculation of the gravitational force exerted on a range of satellite masses placed in orbit just beyond our Lunar orbit at a nominal distance of 400,000 km.

Table 5.7.3 shows a comparative pair of spreadsheet outputs showing that the author's equation digresses from Newton's as the mass of a satellite increases. Note from the author's output column 8 that even with a satellite massing 3,300 tonnes, the ratio of the Force calculated Newton/Roberts is identical to much better than 12 decimal places.

Also, it is proposed that further examination of the author's gravitational equation may produce a clearer view of the significance of outcomes than these illustrative 'snapshot' results.

MG = 1,000,000 kg and distance d fixed at 100 m.
STUDY COMPARISON OF EQUATION OUTPUT OF GRAVITATIONAL FORCE 'F' AS ML INCREASES TO SAME VALUE AS THAT OF MG.

OBSERVATIONS

This scenario shows a rapid divergence of output between the Newton's theory and Roberts' as the mass of the two bodies becomes equal.
In Newton's case, the acceleration of the bodies towards each other stays constant, but in Roberts' case, the acceleration increases with ML's mass increase.
It can be observed that, with an ML mass of 1% or less of MG's, there is very little difference between the Newton and Roberts calculation output.
But, by the time the mass of sphere ML almost equals that of MG, the Roberts value of force is twice the Newton one (ML is 98.526% of MG).
This increase in force for the Roberts calculation as relative sizes become the same reflects the increased intensity of the shadow zone between the bodies.
Note, as shown in Fig 35, the mutual force between the two bodies is controlled by the lesser mass ML.

Newton's 2nd law. f = m.a, where 'a', for example, on the surface of the Earth is approximately 9.81 m/sec2

CALCULATIONS USING NEWTON'S EQUATION $F = G \times MG \times ML2 / d^2$

	G constant N(m/kg)2	MG kg	ML kg	d metres		Force F Newtons	grav. accel (m/sec2)	Ratio of ML/MG as %
0	6.674E-11	1,000,000	1,000	100		6.67E-06	6.67E-09	0.100
1	6.674E-11	1,000,000	1,500	100		1.00E-05	6.67E-09	0.150
2	6.674E-11	1,000,000	2,250	100		1.50E-05	6.67E-09	0.225
3	6.674E-11	1,000,000	3,375	100		2.25E-05	6.67E-09	0.338
4	6.674E-11	1,000,000	5,063	100		3.38E-05	6.67E-09	0.506
5	6.674E-11	1,000,000	7,594	100		5.07E-05	6.67E-09	0.759
6	6.674E-11	1,000,000	11,391	100		7.60E-05	6.67E-09	1.139
7	6.674E-11	1,000,000	17,086	100		1.14E-04	6.67E-09	1.709
8	6.674E-11	1,000,000	25,629	100		1.71E-04	6.67E-09	2.563
9	6.674E-11	1,000,000	38,443	100		2.57E-04	6.67E-09	3.844
10	6.674E-11	1,000,000	57,665	100		3.85E-04	6.67E-09	5.767
11	6.674E-11	1,000,000	86,498	100		5.77E-04	6.67E-09	8.650
12	6.674E-11	1,000,000	129,746	100		8.66E-04	6.67E-09	12.975
13	6.674E-11	1,000,000	194,620	100		1.30E-03	6.67E-09	19.462
14	6.674E-11	1,000,000	291,929	100		1.95E-03	6.67E-09	29.193
15	6.674E-11	1,000,000	437,894	100		2.92E-03	6.67E-09	43.789
16	6.674E-11	1,000,000	656,841	100		4.38E-03	6.67E-09	65.684
17	6.674E-11	1,000,000	985,261	100		6.58E-03	6.67E-09	98.526
18	6.674E-11	1,000,000	1,477,892	100		9.86E-03	6.67E-09	147.789
19	6.674E-11	1,000,000	2,216,838	100		1.48E-02	6.67E-09	221.684
20	6.674E-11	1,000,000	3,325,257	100		2.22E-02	6.67E-09	332.526

CALCULATIONS USING ROBERTS' EQUATION $F = GR \times ML \times (MG + ML) / d^2$

	GR constant N(m/kg)2	MG kg	ML kg	d metres		Force F Newtons	grav. accel (m/sec2)	Ratio of ML/MG as %	Ratio of force F for two equations Newton/Roberts
0	6.674E-11	1,000,000	1,000	100		6.68E-06	6.68E-09	0.100	0.999000999001
1	6.674E-11	1,000,000	1,500	100		1.00E-05	6.68E-09	0.150	0.998502246630
2	6.674E-11	1,000,000	2,250	100		1.51E-05	6.69E-09	0.225	0.997755051135
3	6.674E-11	1,000,000	3,375	100		2.26E-05	6.70E-09	0.338	0.996636352311
4	6.674E-11	1,000,000	5,063	100		3.40E-05	6.71E-09	0.506	0.994962999813
5	6.674E-11	1,000,000	7,594	100		5.11E-05	6.72E-09	0.759	0.992463480445
6	6.674E-11	1,000,000	11,391	100		7.69E-05	6.75E-09	1.139	0.988737660091
7	6.674E-11	1,000,000	17,086	100		1.16E-04	6.79E-09	1.709	0.983201087666
8	6.674E-11	1,000,000	25,629	100		1.75E-04	6.85E-09	2.563	0.975011521132
9	6.674E-11	1,000,000	38,443	100		2.66E-04	6.93E-09	3.844	0.962979820683
10	6.674E-11	1,000,000	57,665	100		4.07E-04	7.06E-09	5.767	0.945478921083
11	6.674E-11	1,000,000	86,498	100		6.27E-04	7.25E-09	8.650	0.920388630504
12	6.674E-11	1,000,000	129,746	100		9.78E-04	7.54E-09	12.975	0.885154451456
13	6.674E-11	1,000,000	194,620	100		1.55E-03	7.97E-09	19.462	0.837086615678
14	6.674E-11	1,000,000	291,929	100		2.52E-03	8.62E-09	29.193	0.774036188176
15	6.674E-11	1,000,000	437,894	100		4.20E-03	9.60E-09	43.789	0.695461609991
16	6.674E-11	1,000,000	656,841	100		7.26E-03	1.11E-08	65.684	0.603558277012
17	6.674E-11	1,000,000	985,261	100		1.31E-02	1.32E-08	98.526	0.503712042085
18	6.674E-11	1,000,000	1,477,892	100		2.44E-02	1.65E-08	147.789	0.403568859504
19	6.674E-11	1,000,000	2,216,838	100		4.76E-02	2.15E-08	221.684	0.310864288453
20	6.674E-11	1,000,000	3,325,257	100		9.60E-02	2.89E-08	332.526	0.231200148894

Table 5.7.4 A comparative set of calculations showing that, in Newton's case, accelerations of the bodies towards each other remain constant as their relative mass varies. In the case of the Roberts equation, the acceleration increases with the mass of ML, until by the time it equals MG it exerts twice the force of the Newtonian equation at the same stage.

151

Table 5.7.4 demonstrate the comparative Newton v Roberts equation outputs for the force between two bodies and their consequent acceleration towards each other as the lesser mass body M_L increases in mass until it is almost the same as the greater mass M_G.

There are several aspects of interest reflected in the comparative outputs of Fig 5.7.4.

For example, using the Newton equation, as the mass of M_L increases, so the force created increases, and thus, the induced acceleration of each body towards the other is kept at a constant figure, based on the standard equation that f=m.a, expressed in this case as a=f/m.

However, in the case of the Roberts equation, as the mass increases, the force between the bodies increases at a higher rate than that of the Newton equation. The outcome is that, by the time the mass of M_L nearly reaches that of M_G, the force generated between the bodies is approximately twice that of the Newton calculation. It can be seen in the Roberts output table that, when the mass of M_L is 98.526% of M_G, Newton's force is only 0.5037 of the Roberts force. In simple terms, at mass parity, Roberts' equation produces twice the force of Newton's.

This fits well with DOPA theory's concept of geometric shielding by the bodies, creating a gravitational 'shadow' zone between them. This is illustrated in Figure 34, which shows that when a smaller body is in proximity to a larger one, there is a smaller physical shadow zone. Figure 30 illustrates the greater intensity of the shadow zone between two equal-sized (and in this case, equal-massed) bodies.

This additional feature of DOPA theory is not present in the Newtonian concept. Distance is the only thing with Newton, but DOPA theory not only decreases its gravitational force between bodies by virtue of the inverse square distance factor, but also by the geometry (and therefore intensity) of the shadow zone as the distance between the bodies changes. The author's exploratory test equation caters for this and produces a reasoned, consequent change in the equation output value of F, which varies accordingly.

The form of the potential force gravity zone, in terms of its dissipation with radial distance from a body, is consequently different from those of both Newton and Einstein in that the DOPA geometrically-generated gravity influence dissipates more rapidly and more realistically than those other two, which purport to extend to infinity.

This section on the Roberts equation has been introduced to indicate to the reader that work is in progress to explore, if not refine and develop, the equation. The author feels that both the numerator and the denominator might be capable of improvement, particularly involving the removal of the inverse d^2 denominator. Also, it is suggested that a definition of where Newton's equation is not valid; for example, 'not within matter' is a first suggestion.

The author invites all readers to experiment for themselves and to write to the author and make any contributions that they wish to be considered. Any accepted contributions will be fully acknowledged in future editions of the book.

5.8 Absorption theory suggests that magnetic material properties may possibly be of indirect help in the study of gravitational force.

The author, very tentatively, suggests it to be possible that materials that have magnetic properties, displaying a unidirectional phenomenon within matter, may have physical properties that would be of interest in research into gravitation—particularly properties that might resist or absorb the passage of gravitational waves in one particular direction.

The development of synthetic materials that permit gravitational waves to pass in one direction but not the other would be the ideal development towards the harnessing and controlling of gravitational force. Magnetic properties may be a natural indicator of this. It is not impossible that a particular lattice geometry within crystalline matter might permit this feature to be developed. This is nothing more than a tentative conjecture at the present moment, but there is no reason why this concept should not be investigated as part of future applied research work.

Any small asymmetrical absorption of gravitational waves by a new synthetic material would provide the world with endless pollution-free, free-of-charge kinetic and thermal energy—a prospect very well worthwhile pursuing.

CHAPTER 6 - WHY OTHER GRAVITATIONAL WAVE THEORIES DO NOT WORK.

There are two alternative theories to DOPA theory. The first is that a planet or star is impermeable to gravitational waves, thus creating a shadow zone between cosmic bodies that drives bodies towards one another by gravitational pressure. The second is that, in addition to gravitational pressure onto a cosmic body, each cosmic body radiates its own gravitational waves. Neither of these proposals can work, as explained below.

6.1 **Simple gravitation wave pressure.**

This concept is the one that was considered for some time before Newton and failed at that time for the same reasons that it would fail today. The proposal concentrates on the idea that, if particles (or gravitational waves) struck the surface of a cosmic body, that would apply a force, thus explaining the pressure that keeps them spherical, and would also explain why there would be a dearth of waves between any two objects, thus driving them together by pressure on their 'outer' surfaces—the famous 'shadow' effect. The shadow part works up to a point, but consideration of individual bodies is where the concept primarily fails.

There are four obvious failings to this proposal.

Firstly, with no outgoing gravitational waves, the incoming waves must always be the same 'background wave flux' force irrespective of the size of the cosmic body. Thus, gravity at the surface of any cosmic body would always be the same, irrespective of its size or density.

Secondly, a pressure concept fails because that is not how gravitational effects are observed in everyday engineering practice and in life generally (See Section 3.10). If gravity were created by pressure alone, that would represent a surcharge environment where we felt pressure on the top of structures or on the top of our heads pushing us downwards.

Thirdly, if matter were opaque to gravitational waves, then we could shield parts of our environment from it using any ordinary material. If the waves come down from space onto the planet's surface and don't penetrate matter (as they do in DOPA theory), then we would only have to walk into a cave or go down a deep mineshaft to feel no gravitational force. Beneath each object, there would be a gravity-free shadow. This does not happen.

Fourthly, surface loading of planets or objects on planets will not provide stress-free acceleration on those objects. Consider the physical stress placed on a car racing driver subject to 1G acceleration or braking. Their body is violently stressed as pressure is either applied by the back of the seat (acceleration) or by the safety harness (braking). However, if you step off the top of a high building and become subject to 1G true gravitational acceleration, you will feel no stress at all as you fall and accelerate, or if the cable holding up your elevator snaps, you will suddenly find yourself in true free-fall and will not feel any stress at all—until you hit the bottom of the lift shaft. That is true gravitational field acceleration, not one induced by pressure. This is a major failing of the pressure concept.

Overall there are just too many obvious failings for this theory to be considered viable, which is why it died out three hundred years ago. Attempts to disinter it are certain to fail.

6.2 **Gravitation wave pressure plus gravitational radiation by matter.**

This theory tries to overcome the failings of wave pressure alone by proposing that all cosmic bodies radiate gravitational waves. This is a step towards DOPA theory, albeit an unsuccessful one. It contains fundamental flaws that DOPA theory does not.

Firstly, we have to consider how the background gravitational flux is created. If it is proposed that it is created by matter creating gravitational waves, then that generates a fundamental problem. Consider the balance of proposed outgoing waves to incoming waves. If the outgoing waves are stronger than or equal to the incoming waves, then all matter at the surface of the body would float or be accelerated away. There would be no 'gravity'. Then, consider the alternative, which is that outgoing gravitational waves are weaker than incoming ones. That would produce a net inward gravitational vector, but it leads to the conundrum of how the universal gravitational flux became stronger than that being emitted by matter. Perhaps it might be claimed that the stronger flux is created by larger, more dense bodies, but that does not work either. If all the stronger emitters are controlling the background level of gravitational force, then the averaged-out level of gravitational force must be somewhere between zero and that maximum level, say half the maximum, assuming an even distribution, or less than half, assuming that there are many more small planets and stars than the bigger ones, which is what we observe. In either case, there will, by the above reasoning, be a large percentage of the planets and stars in the universe that will produce a stronger outgoing gravitational radiation than the incoming radiation strength, which will mean that there will be no inward gravitational force on those planets, and all matter on them will be ejected outwards. This proposal is so fraught with contradictions as to make its failings obvious.

Secondly, we have to consider that, given a fixed (other source) value of background gravitational flux force, the system will break down and not work because of the necessary mechanism for the production of radiated gravitational waves by the cosmic body. It is a necessary assumption that the amount of gravitational energy produced will be proportional in some way to the total mass of the planet or star. In that case (Unlike DOPA theory), the more massive the planet, the greater will be the strength of the outgoing gravitational wave force. Thus, we have the unacceptable conundrum as follows: For a medium-sized planet, we have, say an incoming background gravitational force of 100 m/sec^2 and a counteracting outgoing acceleration force of, say, 50 m'sec^2, resulting in a net inward gravity of 50 m/sec^2. Now, consider a smaller planet that has much less mass, necessarily its gravitational radiation force will be much less. Let's say it has an outgoing force of 30 m/sec2. In that case, the net balance of incoming to outgoing radiated gravitational force is now 100 - 30 = 70 m/sec2 inwards. Its gravity has *increased* because it is smaller. That is entirely wrong and, alone, destroys this theory.

In order to overcome the latter problem, a mechanism would have to be proposed whereby the radiating gravitational strength of a planet would increase as its mass decreases. Impossible to comprehend. That is why this theory fails, while DOPA theory works.

In DOPA theory, the more massive the planet, the more absorption is experienced and the weaker the outgoing gravitational depleted waves are. This means that, with a more massive planet, the outgoing waves provide less resistance to incoming ones and the net inwards force increases. That is what we observe everywhere in the universe. And that is why Roberts' DOPA theory is the only gravitational wave theory that can work.

CHAPTER 7 - CONCLUSIONS

There is much speculation used in attempts by modern physicists to unravel the mysteries of the universe—but, without speculation, we would have no progress.

And yet, the author respectfully proposes that one of the unintentional effects of Einstein's great relativity works was the disincentivisation of research into gravity as a phenomenon in its own right.

At that time, that great genius persuaded cosmologists that we don't live in a three-dimensional universe, we live in a four-dimensional one. He supported the idea that the universe is composed of 'spacetime', and that gravitational force is only a feature of that concept, not something that exists in its own right. And since then, there has not been a single example of research work that has produced a useful demonstration of control or creation of gravitational force.

Consider the enormous number of patents taken out worldwide over the last one hundred years in every branch of science and engineering. Imagine the benefit to humanity in those inventions. And recognise that not one of them has successfully demonstrated the creation, utilisation, or control of gravity.

The author suggests that this is the prime reason why the technology of the universe has proceeded no further. If gravity 'doesn't exist' as laid down by Einstein, then why research it? How can we research something that does not exist? How can we obtain research grants to study and develop something that does not exist?

And so, the question remains: "How can we learn to create, control, and develop gravity?"

We weigh ourselves every so often to see how heavy we are, never thinking about how gravity 'grabs' us and weighs us down. We use gravity in dozens of different ways, without thinking much about it.

The most likely reason that no one studies or researches gravitational force is that no one has, until now, presented any theory proposing how gravitational force is created and, therefore, what it is and, furthermore, proposing convincingly that it can and *does* exist!

That is until now. Because now the author has proposed a new idea in this book as to what could create gravitational force, being a mechanism that we can understand easily, and that does not require amazing new ideas of gravitons and singularities to prop it up. The author hopes to make the general scientific population realise that the perceived concept of spacetime may not only be untrue but may have caused the science of gravitation to die and wither for over a century, bearing no further fruit. A science frozen in time, fossilised and stultified to the impairment of the human race.

It is not important that this particular book should be the last word on the subject, but that it should be the first to open up people's minds to recognise the possibility that even the words of the great

Newton and Einstein must, inevitably, be placed into a newer context—particularly if we are ever to make any further progress in the subject.

It is clear, upon any thoughtful reflection, that the most important field of research and investigation that faces humanity today is not the microscopic behaviour of invisible particles, nor wondering about whether the universe started with a 'big bang', but the science of gravity. We must first realise what gravity is, then investigate it, and by doing so we shall eventually be able to control, if not create, it. The control of gravity will be the greatest scientific development of all time, superseding all technological development of the last ten thousand years. It represents a free source of literally endless, continuous, clean, energy that can be used for the benefit of humanity. Endless clean and controllable power.

It is difficult to come to terms with the fact that this huge potential source of energy has lain dormant, in need of research, for over a hundred years while the rest of science has moved on. It is time that we recognise the futility of our present situation and that gravitation should become a mainstream science and not something that we allow 'not to exist'.

The author, therefore, entreats any and all mathematicians and physicists to read his book and for some of them to commit themselves to quantifying his concepts in order to achieve a state of acceptability that he himself is unable to reach owing to his lack of sufficiently advanced mathematical knowledge. He asks all persons who read this book to feel free to communicate directly with him via his email address at peter.roberts970@gmail.com in order to make one or more contributions or to make such criticisms as they may consider appropriate and to prepare and publish their own peer-reviewed papers on the subject. Particularly, the author would be grateful to be made aware of any mistakes he may have made, and for which he apologises in advance.

And, if you like this book and its contents, then please email the author and tell him, because good testimonials will help him to get his message across to a greater audience.

Peter Roberts
BSc MSc PhD retired Professor
retired CEng CGeol CText FICE FIMMM FIGeol FTI FGS

7th November 2018

APPENDIX 1 - ROBERTS' 1978 PAPER. A THEORY OF GRAVITATIONAL REPULSION

Peter Roberts prepared the following paper in 1978 at a time when computers were, by and large, unknown, and when diagrams had to be hand-drawn. At that time, we knew a lot less about the universe than we do today.

This 1978 paper is offered as a historical document, showing that the seed of the current theory was present at that time, but was lacking the one critical property of differential opposing partial absorption created by the transparency of matter to gravitational waves.

Once the 'penny dropped' and the implications of opposing waves crystallised, then the present theory was born, and this book became possible.

Below is the unaltered paper for the reader's historical interest, so that it may be seen what was lacking in the concept at the time and how much it has changed in the interim.

A THEORY OF GRAVITATIONAL REPULSION

First submitted for publication in July 1978, as witnessed by the 'New Scientist' Editor's acknowledgment letter of 12th September 1978. Copy could be provided if required.

If, instead of each particle in the Universe exerting a mutual attraction, each exerts a mutual repulsion then, as shown in Fig. 1, on any isolated particle situated near the centre of the Universe, all forces would be balanced, and no tendency to move would exist.

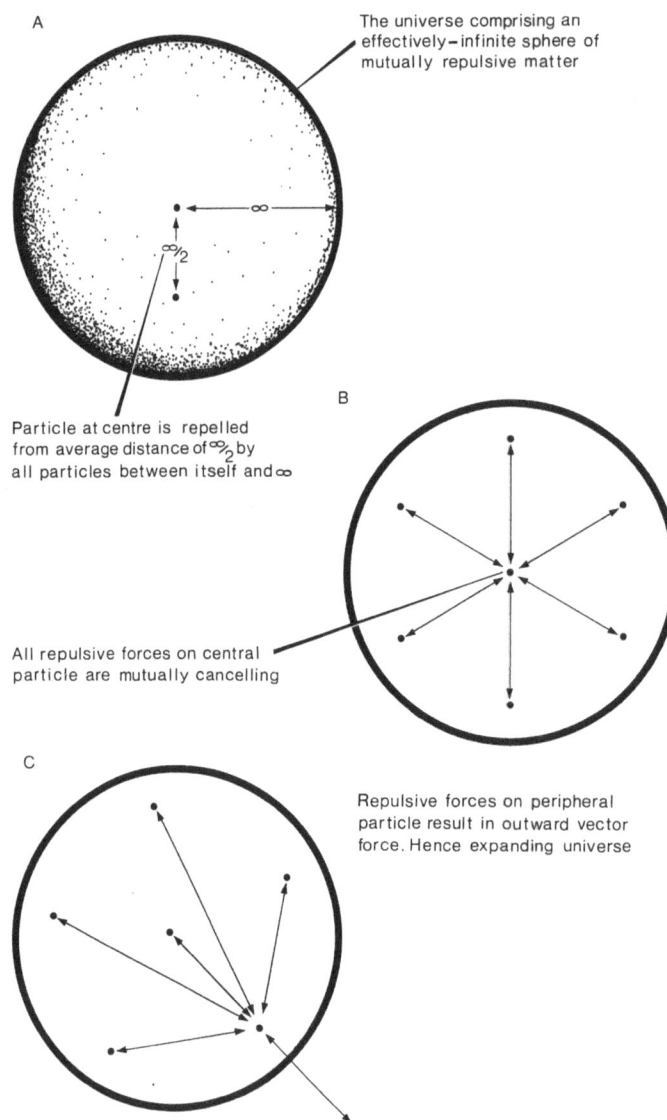

Fig. 1. Stress relationships of particles in a universe with gravitational repulsion.

Consider then (as shown in Fig. 2) two discrete particles a finite distance apart. Each particle now moves towards the other because of the following premises:-

(a) Each particle is acted upon by the forces emanating from the infinite mass of the Universe acting at an effective distance of a half-infinity

(b) The nuclear fields of the constituent atoms of each particle absorb a certain percentage of the gravitational force and thereby shield the other particle from their effect.

(c) Since there is an umbral effect between the two particles a resultant vector force exists upon each particle in a direction towards the other.

(d) As shown in Fig. 3, as the two particles approach each other, their relative shielded area one to another increases, with the effect that the resultant vector force increases in magnitude and the particles are accelerated towards each other at an increasing rate.

(e) Since the distance between the particles is occupied by an umbral cone which blocks off an area of the effective Universal sphere mass, then it becomes apparent why the gravitational force pushing two particles towards each other is inversely proportional to the square of the distance between them.

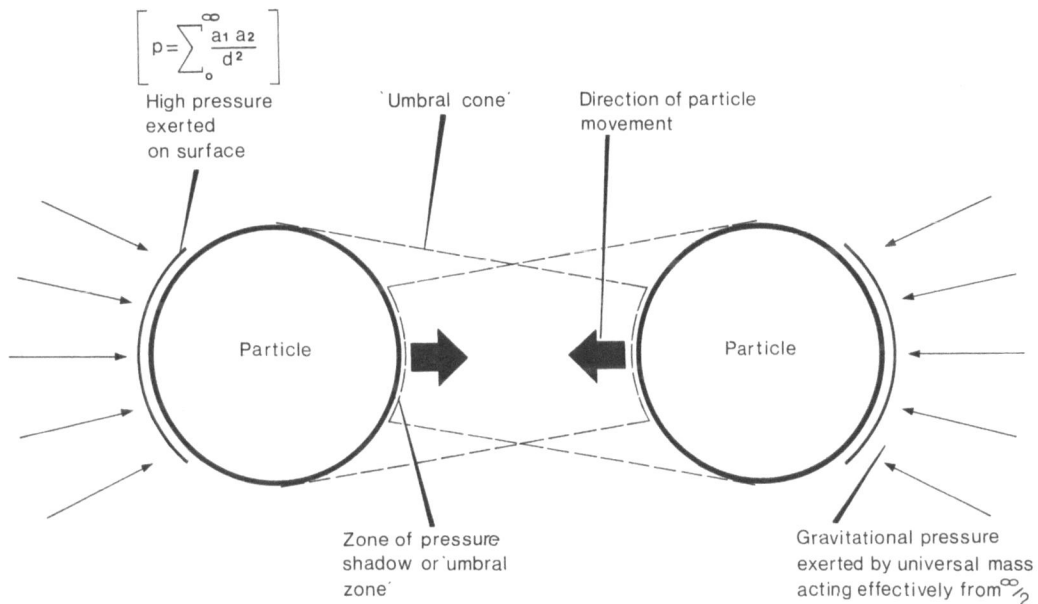

$$\left[p = \sum_{o}^{\infty} \frac{a_1\, a_2}{d^2} \right]$$

High pressure exerted on surface

`Umbral cone`

Direction of particle movement

Particle

Particle

Zone of pressure shadow or `umbral zone`

Gravitational pressure exerted by universal mass acting effectively from $\frac{\infty}{2}$

Fig. 2. Reason why particles move together despite small mutual gravitational repulsion.

The conventional equation governing gravitational attraction, as set out by Newton is:-

$$G = \frac{m_1\, m_2}{d^2}$$

This equation can be paralleled exactly in a repulse theory by the equation

163

$$G = \frac{a_1 \; a_2}{d^2}$$

where a is the absorptivity of the constituent material or any particle.

This concept of gravitational wave absorption allows us to differentiate much more realistically between the effects of true mass (as manifested in the inertia of bodies) and the completely separate effect of bodies being forced together by gravitational pressure.

Similarly, absorption/repulsion theory adequately explains why no apparent gravitational vector force acts upon any particle at the centre of a large body such as the earth, and yet why simultaneously all particles are being forced towards the centre thus generating mechanical pressure.

The Author also points out that the absorption theory would fit together well with the concept of migration of dust particles in space to form accretive bodies. The Author prefers the logic of small particles migrating towards areas of high absorption pushed by the effective force of the Universal mass, instead of the current theory of each tiny particle containing some attractive force pulling itself towards other particles.

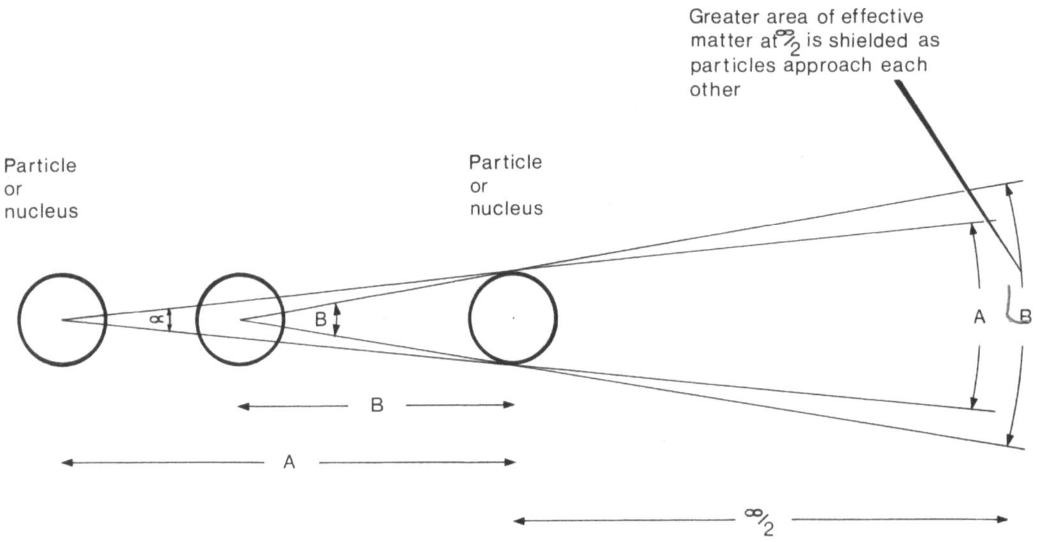

Fig. 3. Increased proximity of bodies causes increased shielding from gravitational pressure on inward-facing surfaces.

This means that throughout space, three-dimensional absorption gradients are established down which particles migrate. It is the absorption gradients themselves that comprise the gravitational field, (the gradients being a secondary product of wave absorption) therefore particle response is instantaneous, and is a reaction to pre-existing gravitational gradients rather than to an inter-particular force.

In a wave-form concept of gravitational repulsion, one must propound that since the size of an atomic nucleus governs our current concept of "weight", then gravitational absorption must be performed by the nucleus with very little effect - if any - from the electron shell. Fig. 4 shows the Author's concept of a gravitational wave passing an atomic nucleus and being partially absorbed by it. In order to generate an effective energy transfer, it is postulated that the wavelength of the gravitational repulsive waves are very short in relation to the diameter of the nucleus.

164

In further consideration of any two particles moving towards each other, it is apparent that they will continue to move together until such a distance is achieved that the repulsive effect of their electro-magnetic fields exactly balances the gravitational force pushing them together.

In the case of individual free atoms or molecules, only a statistical distance can be achieved since energy transfers disturb the inter-particular balance resulting in Brownian movement. In the case of larger bodies, gravitational pressure forces them together until they touch, at which point the electromagnetic repulsion of individual atoms prevents the two bodies passing into each other.

Pressure gravitation adequately fits in with the mechanics of Quasars and Black Holes, and the Author believes that the concept of pressure gravitation explains much more satisfactorily the phenomena of Pulsars and Novae.

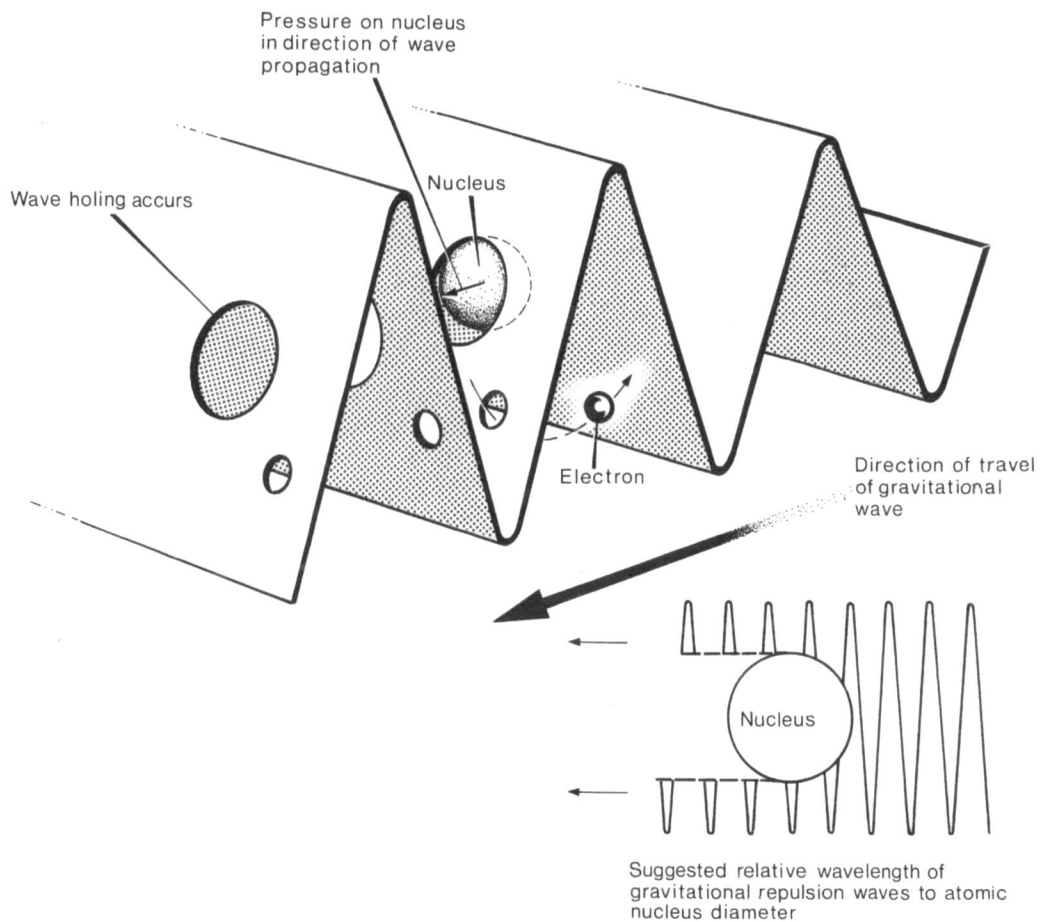

Pressure on nucleus in direction of wave propagation

Nucleus

Wave holing accurs

Electron

Direction of travel of gravitational wave

Nucleus

Suggested relative wavelength of gravitational repulsion waves to atomic nucleus diameter

Fig. 4. Pressure build-up on atomic nucleus by high-frequency gravitational waves.

Under current theory, the short-term gravitational attraction of a star for its own constituent particles should remain fairly constant on the basis that energy can neither be created nor destroyed. Under such circumstances, it is difficult to see why frequent energy oscillations or catastrophic explosions should occur. Under absorption theory, only a small reduction in the size of internal atomic nuclei would be required to reduce their absorptivity and consequently the entire gravitational pressure on the surface of the star. Inherent in this theory is the conservation of energy since the nuclei are still present as before, but are

165

possibly more compact or are changed in some way to reduce their absorption of the force field. Consequent expansion of the star volume would reduce mechanical pressure and thus restore the original absorptivity leading to oscillations (pulsars). Catastrophic explosion (Nova) could result if internal disruption had proceeded too far.

Gravitational Repulsion - Concept of Elastic Deformation of Stress Field.

The concept of gravitational waves being propagated in finite time, and generating gravitational pressure gradients throughout space, leads to a second approach to the envisagement of gravitational pressure as a phenomenon.

The picture of an initially-uniform gravitational field towards the centre of our particular Universe (and distorted by absorption loci) suggests in turn that the overall gravitational field will change towards the periphery of the Universe. Mutual gravitational repulsion would explain most conveniently the current situation of an Exploding (Expanding) Universe and would suggest that the gravitational forces at the periphery would be very much reduced.

If one considers the Universal gravitational field as an elastic medium, then the presence of matter within that medium would create a distortion of the field with resultant stresses. A bubble generated inside an elastic medium sets up spherical stress patterns attempting to reduce the size of that bubble, and containing it. If matter is generated within a gravitational field, then the field may be distorted elastically and attempt to enclose that matter, drive it together, and reduce it to a spherical form, giving optimum re-distribution of stress in the gravitational field, (see Fig. 5A). It is possible that an individual atom would totally distort the surrounding gravitational field, whereas a large body comprised of millions of atoms would only partially distort the field as suggested in Fig. 5B. The tendency to drive all particles together would apply near the centre of the Universe, but towards the peripheries, the effect of repulsion would be to drive all particles apart (thus giving a real effect to any observer present of gravitational repulsion, and providing the mechanism for an expanding Universe).

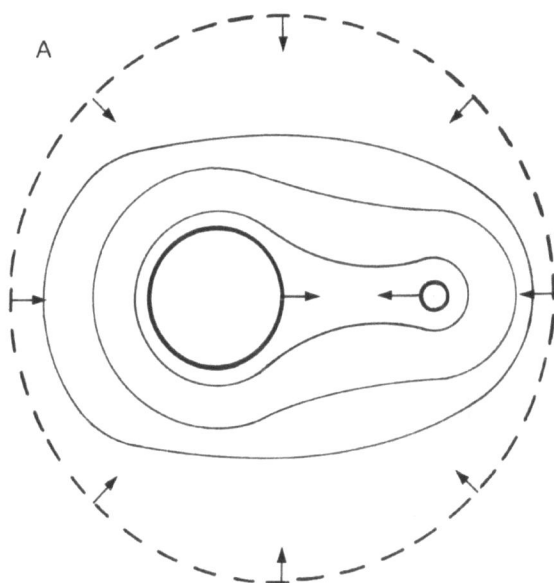

A

Inward pressure caused by the presence of matter straining the universal gravitational field

166

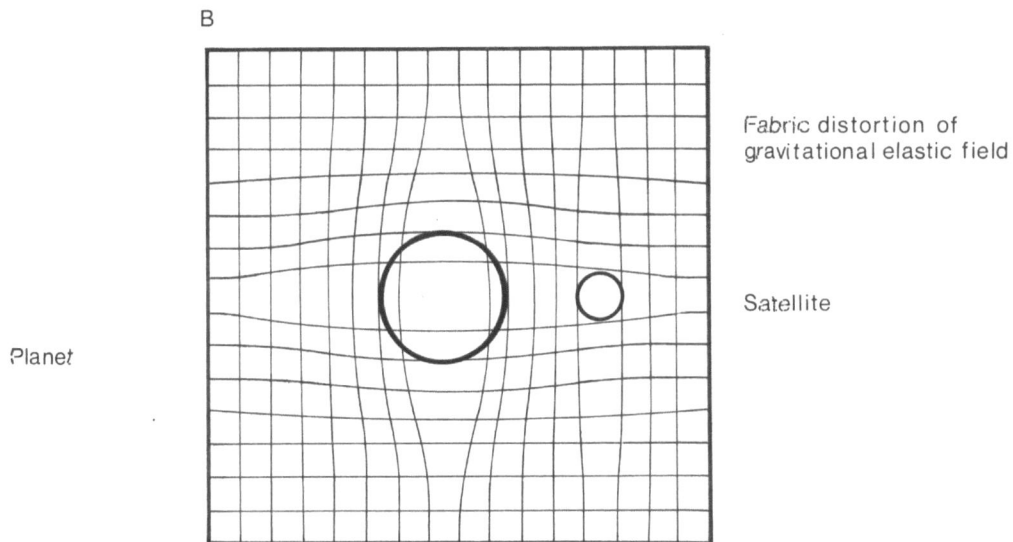

Fig. 5. Distortion of elastic gravitational field by the presence of matter.

Mechanistic model view.

The visualisation of the Universal gravitational field as an elastic medium would also provide a mechanism for an oscillating Universe ranging from total dissipation to contracting nucleation with repeated "Big-Bangs" over a cosmically timed cyclic period. One could see the cycles as follows: -

Expansion (present stage) where all matter is moving away from all other matter and the strain in gravitational field generated by the presence of matter is slowly easing.

1. Maximum Extension   Where the gravitational field approaches zero, matter begins to disintegrate and reconvert to fundamental energy. Energy thus released would be travelling in random directions throughout the Universe.

2. Expansion of the Energy Sphere continues to such a point where the strains within the energy field become negative i.e., tensional, and increase in magnitude until the expansion is slowed down and stage 4 is reached.

3. Elastic Rebound   The energy is elastically drawn back towards the centre of the Universe, and begins to coalesce generating increasingly higher levels of free energy.

4. Matter Generation   The amount of energy present becomes so great that matter begins to "crystallise out" automatically generating repelling forces which grow to such an extent that a minimum size is reached and expansion begins once more, thus returning to stage 1.

The Author acknowledges that the Elastic Field concept is only a mechanistic method of attempting to view the overall proposition of a gravitational absorption/ repulsion theory. He acknowledges that the above paper is descriptive and non-quantitative, and acknowledges that it is intended only to assist in establishing a new angle of thought. After all, it is interesting to think that whether gravity is an attractive or repulsive force, we would not observe any difference.

www.ingramcontent.com/pod-product-compliance
Lightning Source LLC
Chambersburg PA
CBHW041705210326
41598CB00007B/538